How to Build a Time Machine
Paul Davies

PENGUIN BOOKS

Published by the Penguin Group

Penguin Group (USA) Inc., 375 Hudson Street, New York, New York 10014, U.S.A.
Penguin Group (Canada), 90 Eglinton Avenue East, Suite 700, Toronto, Ontario,
Canada M4P 2Y3 (a division of Pearson Penguin Canada Inc.)
Penguin Books Ltd, 80 Strand, London WC2R 0RL, England
Penguin Ireland, 25 St Stephen's Green, Dublin 2, Ireland (a division of Penguin Books Ltd)
Penguin Group (Australia), 250 Camberwell Road, Camberwell, Victoria 3124,
Australia (a division of Pearson Australia Group Pty Ltd)
Penguin Books India Pvt Ltd, 11 Community Centre, Panchsheel Park,
New Delhi – 110 017, India
Penguin Group (NZ), 67 Apollo Drive, Rosedale, North Shore 0632, New Zealand
(a division of Pearson New Zealand Ltd)
Penguin Books (South Africa) (Pty) Ltd, 24 Sturdee Avenue, Rosebank,
Johannesburg 2196, South Africa

Penguin Books Ltd, Registered Offices: 80 Strand, London WC2R 0RL, England

First published in Great Britain by Allen Lane The Penguin Press 2001
First published in the United States of America by Viking Penguin,
a member of Penguin Putnam Inc. 2002
Published in Penguin Books (U.K.) 2002
Published in Penguin Books (U.S.A) 2003

9 10

Grateful acknowledgment is made for permission to reprint an excerpt from
"Burnt Norton" from *Four Quartets* by T. S. Eliot. Copyright 1936 try Harcourt, Inc.
and renewed 1964 by T. S. Eliot. Reprinted by permission of the publisher.

"There was a young lady named Bright" attributed to A. H. Reginald Buller.

Illustrations by Jaye Zimet, adapted from original art by Rebecca Foster and Dan Adams.

THE LIBRARY OF CONGRESS HAS CATALOGED THE HARDCOVER EDITION AS FOLLOWS:
Davies, P.C.W.
How to build a time machine / Paul Davies.
p. cm.
Includes bibliographical references and index.
ISBN 0-670-03063-5 (hc.)
ISBN 978-0-14-200186-8 (pbk.)
1. Space and time. 2. Time travel. I. Title.
QC173.59.S65 D375 2002
530.11—dc21   2002066379

Printed in the United States of America
Set in Univers Light / Designed by Jaye Zimet

Acknowledgments

I am grateful to many people for assisting me with this book. Special thanks are due to colleagues Gerard Milburn, Lee Smolin, Peter Szekeres, Andrew White, and David Wiltshire, as well as my agent, John Brockman, and my editor, Stefan McGrath.

Contents

Contents

Illustrations

A Brief History of Time Travel

Prologue

What if it were possible to build a machine that could transport a human being through time?

Is that credible?

A hundred years ago, few people believed it possible for humans to travel through outer space. Time travel, like space travel, was merely science fiction. Today, spaceflight is almost commonplace. Might time travel one day become commonplace too?

Time travel is inconceivable.
Kingsley Amis

Traveling in time is certainly easy to envisage. You step into the time machine, press a few buttons, and step out again, not just some*where* else, but some*when* else—another time altogether. Writers of science fiction have exploited this theme again and again since H. G. Wells blazed the trail with his famous 1895 story *The Time Machine*. British audiences (the author included) thrilled to the adventures of the time lord Doctor Who and his attractive lady accomplices. Hollywood movies such as *Back to the Future* and books such as *Timeline* make it all seem so easy.

So can it really be done? Is time travel a scientific possibility?

A moment's thought uncovers some tricky questions. Where exactly *are* the past and future? Surely the past has disappeared and cannot be retrieved, while the future hasn't yet come into being. How can a person go to a world that doesn't exist? Sidestepping that, what about the inevitable paradoxes that come from visiting the past and changing it? What does that do to the present? And if time travel were feasible, where *are* all the time tourists from the future, coming back to peer curiously at twenty-first-century society?

There is no doubt that time travel poses some serious problems, even for physicists used to thinking about outlandish concepts like antimatter and black holes. But maybe that is because we are looking at time in the wrong way. After all, our view of time has changed dramatically over the years. In ancient cultures it was associated with process and change, and rooted in the cycles and rhythms of nature. Later, Sir Isaac Newton took a more abstract

> I am afraid I cannot convey the peculiar sensations of time traveling. They are exceedingly unpleasant.
> H. G. Wells

and mechanistic view. "Absolute, true and mathematical time, flowing equably without relation to anything external" was the way he expressed it, and this became the accepted notion among scientists for two hundred years.

Everyone assumed without question that, whatever one's preferred definition, time is the same everywhere and for everybody. In other words, it is absolute and universal. True, we might *feel* time passing differently according to our moods, but time itself is simply time. The purpose of a clock is to circumvent mental distortions and record, objectively, *the* time. Implicit in this view is that time can be chopped up into three parts: past, present, and future. The present—*now*—is supposed to be the fleeting moment of true reality, with the past banished to history—mere shadowy memory—and the future still hazy and unformed. And that all-important *now* is taken to be the same moment throughout the universe: your now and my now are identical wherever we are and whatever we are doing.

Such is the commonsense picture of time, the one we use in daily life. Few people think about time any differently. But it's wrong —deeply and seriously wrong.

That it couldn't be right became apparent about the turn of the twentieth century. The credit for exposing the flaws in the everyday notion of time is largely associated with the name of Albert Einstein and the theory of relativity. At a stroke, Einstein's work demolished Newton's view of both space and time, rendered meaningless the universal division of time into past, present, and future, and paved the way for time travel.

The theory of relativity is nearly a century old. Following publication of the so-called special theory of relativity in 1905, it was accepted by physicists almost immediately. Over the decades it has been exhaustively tested in many experiments. Today, the scientific community is unanimous that "time is relative" and the commonsense notion of an absolute time with a universal "now" is a fiction. Yet among the general public, the relativity of time still comes as something of a shock. Many people seem not to have heard about it at all. Some of them refuse flatly to believe it when told, in spite of the clear experimental evidence.

In the coming chapters we shall see how the theory of relativity implies that a limited form of time travel is certainly possible, while unrestricted time travel—to any epoch, past or future—might just be possible too. If this seems hard to swallow, remind yourself of J. B. S. Haldane's famous dictum: "The universe is not only queerer than we think, it is queerer than we *can* think."

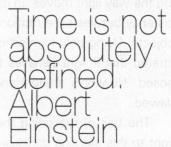

1. How to Visit the Future

In an obvious sense we are all time travelers. Do nothing, and you will be conveyed inexorably into the future at the stately pace of one second per second. But this is of limited interest. A true time traveler needs to leap forward dramatically in time and reach the future *sooner* than everyone else.

Can it be done?

Indeed it can. Scientists have no doubt whatever that it *is* possible to build a time machine to visit the future. And they've known the formula for nearly a century.

Time is not absolutely defined. Albert Einstein

⟋ Time and Motion

It was in 1905 that Albert Einstein first demonstrated the possibility of time travel. He did this by first demolishing the commonsense picture of time dating back to Newton and replacing it with his own concept of *relative* time.

Einstein was twenty-six when he published his special

theory of relativity. He was then not the pipe-smoking disheveled sage with tousled gray hair who provided the role model for many a fictional nutty professor, but a dapper young man in a suit working at the Swiss patent office. In his spare time, the young Einstein was study-

ing the way light moves. In doing so, he noticed an inconsistency between the motion of light and that of material objects. Using only high-school mathematics, he demonstrated that if light behaves the way that physicists supposed, Newton's straightforward idea of time must be flawed.

The trail of reasoning that leads from the motion of light to this startling conclusion about time has been discussed thoroughly elsewhere and need not concern us here. What matters for our purposes is the central claim of the special theory of relativity, which is that

Time is elastic.

It can be stretched and shrunk.

How? Simply by moving very fast.

What precisely do I mean by "stretching time"? Let

me state it more carefully. According to the special theory of relativity, the exact duration of time between two specified events will depend on how the observer is moving. The interval between successive chimes on my clock might be one hour when I am sitting still in my living room, but it will be *less* than one hour if I spend that time moving about.

To express the same thing in a more practical manner, suppose I board an airplane in New York and fly to Rio and back while you stay at Kennedy Airport. Then the duration of the journey according to me isn't the same as the duration according to you. In fact, it is a bit less for me.

Two points need to be made at the outset. First, I'm not talking about the *apparent* duration of the journey. Your experience of being bored at the airport with the hours seeming to drag by, while I am happily occupied watching airline movies, is not the effect being discussed here. Mental time is a fascinating topic in psychology, but my concern is with *physical* time, the sort measured by mindless clocks. The second point is that the time discrepancy for the example given is minuscule—only a few hundred-millionths of a second—far too small to be noticed by a human being; however, it is measurable by modern clocks.

That is pretty much what the physicists Joe Hafele and Richard Keating did in 1971. They put highly accurate atomic clocks into airplanes, flew them around the world, and compared their readings with identical clocks left on

Time ran more slowly in the airplane than in the laboratory.

the ground. The results were unmistakable: time ran more slowly in the airplane than in the laboratory, so that when the experiment was over the airborne clocks were fifty-nine nanoseconds slow relative to the grounded clocks—exactly the amount predicted in Einstein's theory.

Because your time and my time get out of step if we move differently, there can obviously be no universal, absolute time, as Newton assumed. Talk of *the* time is meaningless. The physicist is bound to ask: Whose time?

Significant though the Hafele-Keating experiment may be historically, it is hardly the stuff of science fiction: a timewarp of fifty-nine nanoseconds doesn't make for an adventure. To get a really big effect you have to move very fast. The benchmark here is the speed of light, a dizzying 300,000 kilometers per second. The closer to the speed of light you travel, the bigger the timewarp gets.

Physicists call the slowing of time by motion the time dilation effect. Think of a speed. Divide by the speed of light. Square it. Subtract from 1. Take the square root. The answer is . . . Einstein's time dilation factor! This is a graph of the "slowdown factor." Notice how the graph shows the dilation factor as a function of speed and starts out fairly flat, but plummets as light speed is approached. At half the speed of light, time is about 13 percent slowed; at 99 percent, it is seven times slower—1 minute is reduced to about 8.5 seconds.

Technically, the timewarp becomes infinite when the speed of light is reached. This is a sign of trouble. In fact,

The answer is . . . Einstein's time dilation factor! This is a graph of the "slowdown factor."

it tells us that a normal material body can't reach the speed of light. There is a "light barrier" that can never be breached. The no-faster-than-light rule is a key result of the theory of relativity:

Nothing can break the light barrier.

This includes not just material bodies but waves, field disturbances—physical influences of any sort. It spoils a lot of science fiction because, fast though it goes, light still takes a long time to cover interstellar distances. The nearest star, for example, is over four light-years away, which means it takes light over four years to get there from Earth. The Milky Way galaxy is about 100,000 light-years across. Administering a galactic empire would be a slow process.

However, there is some compensation. Because time is stretched by speed, interstellar journeys would seem quicker for the astronauts than for those left on Earth at mission control. In a spaceship traveling at 99 percent of the speed of light, a trip across the galaxy would be completed in only 14,000 years. At 99.99 percent of the speed of light, the gain is even more spectacular: the trip lasts a mere 1,400 years. If you could reach 99.999999 percent of the speed of light, the trip could be completed in a human lifetime.

Speeds like this are far beyond current spacecraft technology. (Our best spacecraft reach a paltry 0.01 percent of the speed of light.) But there are objects that travel very close to the speed of light. These are subatomic particles, such as cosmic rays and atomic fragments emitted in radioactive decays, or purposely accelerated in giant "atom smashers." It's possible to observe very large time dilations by using these particles as simple clocks. The particle accelerator known as the Large Electron Positron (LEP) collider at the Centre Européenne pour la Recherche Nucléaire (CERN) laboratory near Geneva could propel electrons to 99.999999999 percent of the speed of light. This is so fast it falls short of the speed of light by a literal snail's pace. At this speed, timewarp factors approaching a million were achieved. Even this pales into insignificance conpared to timewarp factors of billions experienced by some cosmic ray particles.

In a series of careful experiments carried out at CERN in 1966, particles called *muons* were circulated inside a small accelerator to test Einstein's time dilation equation to high precision. Muons are unstable and decay with a known half-life. A muon sitting on your desktop would decay on average in about two microseconds. But when muons were moving inside the accelerator at 99.7 percent of the speed of light, their average lifetime was extended by a factor of twelve.

The effect of motion on time is often discussed using the parable of the twins. It goes something like this. Sally and Sam decide to test Einstein's theory, so Sally boards a rocket ship in 2001 and zooms off at 99 percent of the speed of light to a nearby star situated ten light-years away. Sam stays at home. On reaching her destination, Sally immediately turns around and heads home at the same speed. Sam observes the duration of her journey to be just over twenty Earth years. But Sally experiences time differently. For her, the journey has taken less than three years, so when she gets back to Earth she finds that the date there is 2021 and Sam is now seventeen years older than she is. Sally and Sam are no longer twins of the same age. In effect, Sally has been transported seventeen years into Sam's future. With a high enough speed, you could "jump" to any date in the future you like. The year 3000 could be reached in less than six months by traveling at 99.99999 percent of the speed of light.

Traveling through time works the opposite way from traveling through space. The shortest distance between two points is a straight line, so in daily life you get from A to B most quickly by following a direct route. But when it comes to time travel, it is stay-at-home Sam who ages more; that is, he takes longer to reach year 2021. By zooming about, Sally dramatically shortens the time differ-ence between the two events "Earth year 2001" and

"Earth year 2021." In fact, the more she zooms this way and that, the shorter the time difference between these two events becomes.

Some people find the twins effect paradoxical, because from Sally's point of view, she is at rest in the rocket ship while the Earth flies away. However, there is no paradox, because the situation for Sally and Sam is not symmetrical. Sally is the one who accelerates away by firing the rocket motors, then maneuvers around the star, and finally decelerates to land on Earth. These changes in motion single her out as the one to age less.

Note that Sally cannot "get back" to Earth year 2007 (there being six years' round-trip travel time after departure) this way, in order to reequalize her age with Sam's. If she reverses her trajectory, she will succeed only in leaping another seventeen years into Sam's future. High-speed motion is a one-way journey into the future.

⊘ How to Use Gravity to Travel into the Future

Speed is only one method of warping time. Another is gravity. As early as 1908 Einstein began extending his special theory of relativity to include the effects of gravity. Using another ingenious argument concerning light, he came to the remarkable conclusion that

Gravity slows time.

He didn't clinch the argument until 1915, when he presented his so-called general theory of relativity. This work extended the special theory published in 1905 to include the effects of gravitational fields on time, and on space too.

Putting the numbers into Einstein's equation shows that the Earth's gravity causes clocks to lose one microsecond every three hundred years. This leads to the curious prediction that

Time runs faster in space.

But not so much that astronauts notice. (You would gain just a couple of milliseconds by spending six months aboard the International Space Station.) However, physicists can readily measure the effect using accurate clocks. In 1976, Robert Vessot and Martin Levine flew a hydrogen maser clock into space from West Virginia and monitored it carefully from the ground. Sure enough, the rocket-borne clock gained about one-tenth of a microsecond before crashing into the Atlantic Ocean a couple of hours later.

There is even a tiny time difference between the bottom and top of a building. In 1959 an experiment was carried out at Harvard University to measure the timewarp factor up a tower 22.5 meters high. A slowing effect of

The rocket-borne clock gained about one-tenth of a microsecond before crashing into the Atlantic Ocean.

0.000000000000257 percent was detected, by using an extremely accurate nuclear process. Small it may be, but the measured value confirmed Einstein's prediction. Nobody was really surprised at this result, as physicists had long accepted gravity's effect on time.

If you could magically squash the Earth to half its diameter (retaining all its mass), its surface gravity would be twice as big, and so would be the timewarp. Go on compressing, and the effect rises. When the radius reaches a critical value of 0.9 cm, time "stands still." Nothing can escape! The graph shows the "slowdown factor" for a clock on the surface of the contracting ball. Notice how the timewarp becomes infinite when the ball is shrunk to about the size of a pea.

Of course, squashing all that matter into a cubic centimeter is a pretty fanciful notion. But stupendous compressions do occur in astrophysics. For example, when stars run out of fuel they shrink spectacularly under their own weight, ending up a tiny fraction of their original size. Some large stars actually implode, quite suddenly, and form spinning balls not much bigger than Manhattan, yet containing masses greater than the Sun (about two thousand trillion trillion tons). The gravity of these collapsed stars is so great that even their atoms are crushed to form neutrons, so they are known as "neutron stars." One such object lies in the constellation of Taurus, deep within a ragged cloud of expanding gas

When the radius reaches a critical value of 0.9 cm, time "stands still." Nothing can escape!

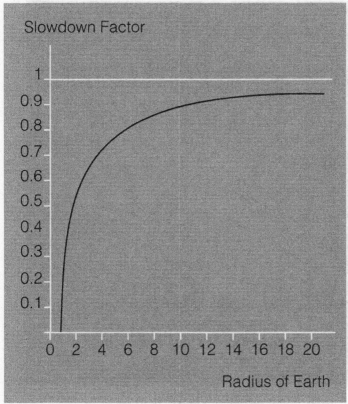

The graph shows the "slowdown factor" for a clock on the surface of the contracting ball.

The Crab Nebula.

called the Crab Nebula. The nebula contains the shattered remains of a giant star that was seen to explode in 1054 by Chinese chroniclers.

Astronomers have discovered many more such objects and determined that the gravity at their surfaces is large enough to cause substantial timewarps. A clock on a typical neutron star would tick about 30 percent slower than one on Earth. So take up residence near a neutron star (admittedly not a very practical proposition), and you have a ready-made time machine for journeying into the future. Seven years spent there would correspond to ten years passing by on Earth.

If you could look back at Earth from the surface of a neutron star, you would see terrestrial events speeded up, like a fast-forward video show. Events in your immediate vicinity would seem normal, though. It wouldn't *feel* as if you were living in a high-speed world, or that mental time was disconcertingly whizzing by.

Is all this true? Yes, it is. There are a pair of neutron stars in the constellation of Aquila that cavort about each other, emitting regular radio bleeps, enabling astronomers to confirm with great precision the timewarping effects that Einstein's general theory of relativity predicts.

$\mathcal{O}$ Is It Really Time That Slows?

Some people object that the theory of relativity merely describes how *clocks* are affected by motion and gravitation, not time itself. This is a misunderstanding. Clocks measure time. If all clocks (including the human brain, which governs our personal perception of time) are slowed equally, then it is correct to say that time itself has slowed, for there is no duration of time other than what can be measured by clocks (of some sort). Similarly, if all distances were shrunk in length by the same factor, it would be true to say that space had shrunk.

To make this point clear, suppose I have an aging and delicate grandfather clock that I put on a jet plane to test the time dilation effect. If the clock falls to bits as the plane roars down the runway, it would be wrong to conclude that time stands still on board the plane because the clock is no longer ticking. To make sense of time dilation, the effects of acceleration on the clock mechanism must be factored out before concluding anything about time itself. Time dilation is the pure time phenomenon that remains. Note that during smooth motion, such as uniform flight in an airplane, there are no mechanical effects on clocks anyway. (Galileo long ago taught us that uniform motion is only relative.) A constant velocity does not lead to any forces that would affect a clock; otherwise, we'd have to worry about how the clock would depend on the speed of Earth through space.

$\bigcirc$ $E = mc^2$: Einstein's Famous Equation

Even those with no scientific education will be familiar with Einstein's famous equation $E = mc^2$. It will play a crucial role in the discussion of time travel. The symbols here stand for energy, E; mass, m; and the speed of light, c. The theory tells us that mass and energy are related; that is, energy has mass and mass is a form of energy. In the diagram the swinging pendulum is very slightly heavier than the static one, all else being equal, because the kinetic energy of the pendulum has mass. The conversion factor c^2 is a very big number because the speed of light is so great. This means a little bit of mass is worth an awful lot of energy. For example, one gram of matter, converted into electricity, could power an entire city for several days. Nuclear reactions of the sort used in power stations convert about 1 percent of the mass of the fuel into energy, a much higher yield than chemical reactions. Conversely, familiar quantities of energy don't have much mass. The heat energy needed to boil a kettle dry would weigh a measly fifty picograms.

Energy enters the time machine story via gravitation. Mass is a source of gravity. As energy has mass, it must gravitate too. The heat energy inside the Earth, for example, contributes a few nanograms to your body weight.

Einstein derived his equation from the special theory of relativity. One way to glimpse the link is to reflect on the fact that material bodies cannot go faster than light. So

The swinging pendulum is very slightly heavier than the static one.

what happens if you just go ahead and try to accelerate a particle of matter through the light barrier? This is precisely the sort of thing that physicists working with subatomic particles do with their giant accelerator machines. The result is that as the particle gets nearer the speed of light, it becomes heavier—that is, puts on mass. (An electron whirling around inside the LEP accelerator, for example, weighed about 200,000 times an electron at rest.) This makes the particle harder and harder to speed up. More and more of the energy goes to making the particle heavier, less and less to increasing its speed. The speed of light is the final barrier; if the particle could get there, this would imply that its mass is infinite. To make it go any faster would therefore require an infinite force, which is impossible.

⊘ The Future Is Out There

Although he wrote ten years before Einstein's special theory of relativity, H. G. Wells realized that time could be thought of as the fourth dimension. He surmised that just as we can move through the three dimensions of space, so it might be possible to move through the time dimension too. But this beguiling idea tacitly assumes that the past and future are "out there" somewhere, so it's not merely the present that is real. Physicists do indeed think of all time as equally existent—making up an extended "timescape." To be sure, the concepts of past, present, and future are convenient linguistic devices in the realm of human affairs, but they have no absolute physical significance. Einstein himself expressed it bluntly in a letter to a friend. "The distinction between past, present and future," he wrote, "is only an illusion, even if a stubborn one."

This often strikes nonphysicists as crazy. How can the past and future exist alongside the present? Einstein gave the following argument for why we can't dissect time neatly into past, present, and future in a way that all observers would agree on. Start by asking: How do we know that "now" in one place is the same as "now" in another? Think about this. Suppose it is 6 P.M. where you are. What events are happening on the other side of the world at the same moment? Einstein insisted that there was no proper answer to such a simple question.

Why, you might wonder? Can't we just phone somebody and do a blow-by-blow comparison? Well, the problem is that it takes time for telephone signals to travel, even at the speed of light. In fact, it takes about seven-hundredths of a second for voice messages to traverse the globe in optical fibers. (The delay is not quite noticeable to the human ear.) So the news from the other side of the world always arrives a bit late. (Not much, granted, but I am making a point of principle.) If your friend was on Mars, you might wait twenty minutes to learn what was happening. Since it is a fundamental principle of physics that no signal can travel faster than light, some delay is inevitable.

In itself, the delay is no problem in trying to establish simultaneity; you could simply compensate by subtracting the requisite time interval needed for the signal to arrive. The real difficulty lies in the fact that observers who move differently disagree on the value of this compensating factor. That is because their clocks tick differently, owing to the time dilation effect. So opinions will differ, depending on whom you consult, on how much delay has elapsed while light (or radio) signals are traveling between A and B. An astronaut rushing past Earth at half the speed of light would be seriously at odds with an earthbound observer in deciding on the precise delay time for a round-the-world signal.

As a result of such mismatches, there is *no* unique event on the other side of the world, or on Mars, or generally at any other point in space apart from where you are located, that is exactly simultaneous with your "now." There will be a *range* of such events at distant places. Which particular event is judged to be happening at the same moment as "6 P.M. at home" will depend on just how the observer is moving. The ambiguity isn't much when restricted to Earth (just a fraction of a second this way or that), but the range of contending nows grows with distance. For Mars it is some minutes. For a star on the other side of the galaxy, events happening at the same moment as an event on Earth today might lie anywhere in a time span of 100,000 years.

The upshot is there can't be a single present moment that is the same for everybody everywhere. To spell it out:

There is no universal "now."

We have to accept that time at a faraway place must extend somewhat into our perceived past and future. And by symmetry, distant observers will regard time on Earth as extending into their past and future. There is no other way to make sense of the facts. Obviously, then, it's wrong to think of only the present as real, right across the cosmos. Some events that you judge to be in the past will be regarded by someone else as lying in his or her future or present—and vice versa.

To take a concrete example, Earth has a definite history, and so does a hypothetical Planet X situated 5,000 light-years away. Attempts to compare dates of specific events on the two planets are pointless because the alignment of the respective timelines is ambiguous over a span of thousands of years.

This *doesn't* imply that the order of cause and effect can be reversed simply by traveling fast. Let me explain why. Events have an ambiguous time order only if light doesn't have long enough to pass between them. For example, if I fire a gun on Earth and an astronaut fires a gun on Mars one second later (by my reckoning), an observer in a speeding rocket ship might well judge the Mars gun

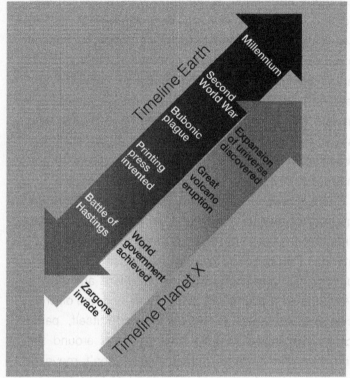

Attempts to compare
dates of specific
events on the two
planets are pointless.

as having been discharged first. But if the Mars gun goes off a week after mine, everyone agrees on which was fired first, as a week is easily long enough for light to travel between Earth and Mars. If no physical influence can exceed the speed of light, ambiguously time-ordered events can never affect each other, so causality isn't threatened. Notice, however, that if the no-faster-than-light rule was wrong, causality would be in trouble, and past and future would become jumbled. As we shall see, this little clue will turn out to be highly significant for the construction of a general-purpose time machine.

There is never any ambiguity about the time order of a sequence of events happening at one place; nobody claims that the battle of Hastings came after the battle of Waterloo. The quibbling comes only when events here and now are compared with events *there* and now—where "there" is a long way away. Even then, the discrepancies are too small to notice on Earth itself, partly because light takes so little time to travel around the world, but also because human beings don't move at more than a tiny fraction of the speed of light anyway. However, that is incidental. The crucial point is that there can be no absolute meaning assigned to "the same moment" at two different places.

So the future is out there all right, and it *can* be visited. All you need as an effective time machine is a spaceship that can travel very close to the speed of light or withstand the lethal conditions near a neutron star. Ultrahigh speed is not a problem in principle, merely a practical difficulty that may be overcome someday. The major drawback is the energy cost. To accelerate a ten-ton payload to 99.9 percent of the speed of light requires an energy expenditure of ten billion billion joules, equivalent to humanity's entire power output for several months. And the energy needed grows in direct proportion to the time-warp factor: halving the clock rate demands twice the energy. With these costs, nobody is going to take a great leap forward in time using rocket technology. If a way could be found to tap natural sources of energy in space, near-light travel might one day be achievable. Then the future would lie within our grasp.

What about coming back from the future?

High-speed travel and gravitational time dilation can be used only to go forward in time. But just as the future is surely out there, so is the past. It's there for the visiting. The trick is to figure out a way to reach it.

2. How to Visit the Past

The first hint that certain gravitational fields might permit travel *backward* as well as forward in time came in 1937, with a little-known paper by W. J. van Stockum published in a Scottish scientific journal. Van Stockum used Ein-

There was a young lady named Bright
Whose speed was far faster
  than light;
  She set out one day
  In a relative way
And returned on the previous night.
Punch, December 19, 1923

stein's general theory of relativity to predict what would happen if an observer went into orbit around a rotating cylinder. He found that if the cylinder spun fast enough, the observer could return to her starting point *before* she left. In other words, a closed loop in space could also become a loop in time. Nobody got excited because, to simplify the mathematics, van Stockum had assumed, unrealistically, that the cylinder was infinitely long. Nevertheless, this result served to show that the general theory of relativity did not explicitly forbid travel into the past. Another fifty years were to pass before physicists found a more realistic way to make a time machine.

How to Travel Faster Than Light

A decade after publication of van Stockum's paper, the eminent Austrian logician Kurt Gödel produced another solution of Einstein's equations of general relativity that contained time loops. Gödel was then working at Princeton's Institute for Advanced Study alongside Einstein. He discovered that if the entire universe were rotating, it would be possible to find orbits in space that spiraled back into the past. In fact, Gödel showed that in such a universe you could depart from Earth and travel to any*where* and any*when* you liked.

Gödel's mathematical model was intended as a curiosity only, not as a serious proposal. Even in the 1940s astronomers had good reason to doubt that the universe as a whole is spinning, although individual galaxies are. Today, measurements of the heat radiation left over from the

big bang can be used to determine with great accuracy any cosmic rotation, and none is observed. Despite its manifestly artificial nature, Gödel's model universe seriously disturbed Einstein, who admitted that he'd worried about backward time travel ever since he first formulated the general theory of relativity.

What is the secret that lets rotation open a gateway to the past? A pointer lies in the famous limerick at the start of this chapter. As I explained in chapter 1 (see p. 28), given the no-faster-than-light rule, the time order of events that can be connected by light signals is never in doubt. But if faster-than-light motion were to be allowed, causal chaos would ensue. It would be possible to reverse cause and effect or, to put it another way, to move in such a way as to change "before" and "after" for some pairs of spatially separated events. It is but a small step from this reversal of time order actually to visit the past. In other words, "faster than light" can mean "backward in time."

But there is a subtlety. Rotation does not enable an astronaut, or even a particle of matter, to break the light barrier as such, but does affect the motion of light itself. According to the general theory of relativity, if a massive body (for example, a cylinder, or the universe) is spinning, it will act like a vortex in space and drag any passing light beam around with it. Normally, this dragging phenomenon is tiny, but if the body is heavy enough and spins fast enough, light can be caught up in the twisted gravitational field and pulled right around in a loop. If an intrepid astronaut ventures into this gravitational whirlpool, he or she, too, will be caught up and dragged around. At all times, the astronaut travels around the spinning body *more slowly* than the light in her vicinity. But because the light itself is being whirled around, in effect the astronaut is achieving Miss Bright's status relative to a distant observer. Locally, the light barrier isn't broken, but globally—considering the entire circuit—the astronaut seems to reach superluminal speed. In the 1970s, the physicist Frank Tipler showed that a superdense cylinder spinning on its axis at half the speed of light could serve in this manner as a time machine, although the senario he outlined was not physically realistic.

Although the recent ideas for time machines don't require rotation, they, too, involve a way to effectively outpace light. The most popular proposal is the "wormhole." As we shall see, this is a sculpture in the structure of space that provides a shortcut between two widely separated places. By traveling through the wormhole, an astronaut would be able to go from A to B before light had had a chance to get there the long way, that is, across normal space.

So what, exactly, is a wormhole? To introduce the concept, I must first explain that better-known object, the black hole.

⟋ How to Make a Black Hole

Black holes are certainly newsworthy, and most people are now familiar with the basic idea: dense, dark bodies in space that suck in everything around them. Small black holes a few kilometers wide form when large stars burn out and collapse under their own weight. Some become neutron stars, others black holes. Our Sun is likely to escape either fate—being of fairly modest mass—and will probably end its days as a white dwarf. Some astronomers think the galaxy could be peppered with myriad black holes, the dead remnants of giant stars born billions of years before the solar system.

Much larger black holes lurk at the centers of galaxies. Our own Milky Way seems to harbor an object with the equivalent mass of about a million suns. Other galaxies are known that host central black holes one thousand times bigger still. Sometimes material spiraling into these supermassive objects releases vast quantities of energy, creating violent disturbances. Intense radiation is given off, and jets of material spew forth at near the speed of light.

Why the term "black holes"? The name was coined by the physicist John Wheeler in the late 1960s. He chose it carefully to encapsulate two defining properties: blackness and emptiness. A neutron star contains what is probably the stiffest material in the universe, but even that is not completely incompressible. If it could be further squeezed, the pull of gravity would become overwhelming, and the star would collapse completely. That happens inside large stars that run out of fuel and can no longer sustain their internal pressure. The core abruptly implodes, in a fraction of a second, leaving behind a region of empty space—hence, "hole." (Well, in practice the surrounding region isn't totally empty because of the ragged remains of the rest of the star. But that stuff soon either gets blown away or sucked in.)

Mystery surrounds the fate of the imploding matter. Where does it go? For now, think of the star as a precisely spherical ball. Imagine it being progressively shrunk, with all the matter retained. As I explained on p. 17, the smaller the ball gets, the higher the gravity becomes at the surface. At some point gravity would be so great that no known matter could withstand it, and the ball would collapse. Because it is an exact sphere, and nothing disturbs the symmetry, the ball

must remain a sphere as it implodes. In other words, all the material must move toward the precise geometrical center. The smaller the sphere gets, the more powerful is the pull of gravity inside it and the faster it shrinks.

Where does it all end? Under these conditions it can end only with the entire contents of the ball concentrated at a single point at the center. Obviously, this dot of matter would have infinite density, and its gravity would also be infinite there. Mathematicians refer to such an entity as a *singularity*. When infinity looms in a physical theory, it is an alarm signal, suggesting that something drastic happens, but in this case nobody is quite sure what. Shortly I shall have more to say about singularities, but for now it is enough to note that whatever is the ultimate fate of the ball, it does not affect the gravitational field outside it. The ball's gravity doesn't go away just because it has imploded. Like the fading grin of the Cheshire cat, the ball's erstwhile existence leaves an imprint in the surrounding universe in the form of its ferocious gravitational field.

Let me now turn to the other property that characterizes black holes—their blackness.

In the first chapter I explained how gravity slows time; the stronger the gravity, the bigger the timewarp. Think what happens to time at the surface of the ball as it contracts. The slowdown factor rises as the radius gets smaller. When the ball approaches a certain critical radius—about three kilometers for an object containing one solar mass—the timewarp becomes infinite, which is to say that the march of time at the surface of the ball grinds to a halt, relative to, say, Earth time. A clock on the surface of the ball would appear from afar to be frozen into total immobility.

Of course, no manmade clock could withstand the huge forces involved here, but light waves can be regarded as a type of clock; their undulations mimic the swing of a pendulum. So the light from a shrinking star gets lower and lower in frequency as the escalating timewarp retards its oscillations. Translated into color, the light from the contracting ball gets redder and redder, until it fades away completely, like the dying embers of a cooling fire. Eventually, the last of the light from the star gets out; after that, all is black. Thus, the region of space around the collapsed object is both black and empty—hence, black hole.

The way I've described it makes the ball's disappearance as viewed from Earth—by our twin Sally, say—seem rather slow going. In fact, for a star containing the equivalent of one solar mass, the fade-out time is as short as a few hundredths of a millisecond. Sally would see the core of the star vanish in an instant (assuming she could see the core anyway), and the region of space where the ball of matter used to be would be occupied by a featureless black sphere—a black hole.

An observer—Sam, let's say—standing on the surface of the contracting star, and unlucky enough to accompany it into the black hole, would experience events very differently. No temporal slowdown for him. Remember, time is relative. In fact, in this case the two accounts given by Sally and Sam are *infinitely* different, because of the infinite timewarp. Sally sees the star collapse to a three-kilometer black ball and freeze—permanently—while Sam sees the entire star shrink to nothing in the twinkling of an eye. As far as Sam is concerned, in the fraction of a second it takes the star to implode across the critical radius, all of eternity will have passed by in the outside universe.

The formation of an *infinite* timewarp around an imploding ball of matter leads to an arresting conclusion:

A black hole is a one-way journey to nowhere.

You can't fall into one and come out again, because the region inside the black hole is *beyond the end of time* as far as the outside universe is concerned. If you did somehow manage to emerge from a black hole, you would have to come out before you fell in. Which is another way of saying that you would be projected back in time.

So there's a clue here. A black hole has an entrance but no exit; it is a one-way fast track to the end of time.

What if there existed something like a black hole, but with an exit as well as an entrance—a *wormhole*? Maybe it could be used to reach the past.

⊘ Wormholes and Curved Space

In order to explain what a wormhole is I need to describe how gravity affects space as well as time. The theory of relativity demands that both space and time be elastic. That means space can stretch too; in fact, the expansion of the universe is more or less just that. Because space has three dimensions, however, its elasticity can produce a broader range of distortions than simple stretching and shrinking; space can also be *curved*.

So what does curved space mean? At school we learn the rules of geometry compiled by Euclid. To give a simple example, the three angles of a triangle add up to two right angles (180 degrees). Euclid's rules apply to geometrical shapes drawn on blackboards and in exercise books, which are flat. But on a curved surface, the rules of geometry are different. For example, on a spherical surface such as the Earth it is possible to draw a triangle with three right angles (270 degrees). The apex of the triangle is at the North Pole and the opposite side lies along the equator. Navigators are familiar with the fact that on the Earth's surface different rules of geometry are needed. Similar rules can apply in three-dimensional space if it is appropriately warped.

To give an example, imagine drawing a *flat* triangle around the Sun. What would the angles add up to? Most people would guess 180 degrees. The theory of relativity predicts that the answer should be a little bit more than 180 degrees, because the Sun's gravity curves the space around it. The effect is very small—a few seconds of arc for a triangle that just encloses the Sun, still less for a bigger one. Nevertheless, the distortion can be measured,

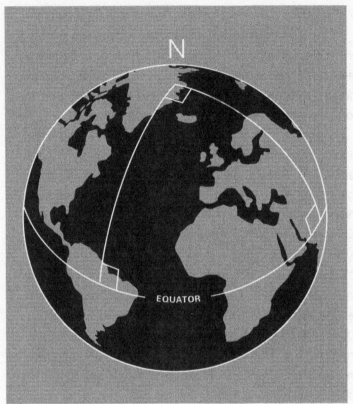

On a spherical surface
such as the Earth it is
possible to draw a
triangle with three right
angles.

not by literally drawing a triangle, of course, but by observing light rays or radar signals passing close to the Sun. Sometimes the effect is described by saying that the Sun's gravity bends light rays, but it is more accurate to think of space itself as bent, with light following the shortest path through the curved geometry.

The curvature of space around the Sun is barely discernible. A really big spacewarp demands a much stronger gravitational field, such as that of an entire galaxy containing hundreds of billions of stars. Sometimes by chance one galaxy will line up exactly in front of another as seen from Earth. In this case the intervening galaxy's gravitational field serves as a type of lens, bending the light from the more distant galaxy and focusing it, producing a halo effect known as an Einstein ring. At the surface of a black hole with the mass of one sun, gravity is about one hundred billion times stronger than at the surface of the Sun, and space is spectacularly warped.

One way to picture elastic space near a massive object is by analogy with a rubber sheet. The sheet is laid horizontally with a pit in the middle made by placing a heavy ball there. This might represent the Sun's spacewarp, for example. A smaller ball rolled across the sheet will, in negotiating its way across the warped surface, travel in a curved path around the pit, just as the Earth travels in a curved orbit around the Sun. Of course, the sheet here represents only two space dimensions; in reality the Sun's gravity curves space in all *three* dimensions,

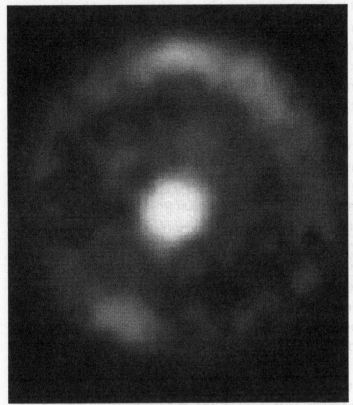

A halo effect known as
an Einstein ring.

but that is hard to show in a picture. (Strictly speaking, it warps all *four* dimensions of spacetime—see below.)

Imagine that instead of the Sun, there is a black hole. The rubber now curves away dramatically, into an apparently bottomless pit. A lot of thought has been given to what lies at the bottom of the pit, or whether there is any bottom at all. It may end in a so-called singularity—an edge of spacetime. As early as 1916 the Austrian physicist Ludwig Flamm studied the geometry of space in and around what we would now call a black hole, although that term wasn't coined until 1968.

Later, in 1935, Einstein and his collaborator Nathan Rosen revisited this topic, and the shape shown on page 47 became known as the Einstein-Rosen bridge. Today, a structure of this general form is called a wormhole, and the narrow bit in the middle is termed the "throat." Far from the black hole the sheet is nearly flat, because gravity is weak there. As the hole is approached, the curvature rises, and the sheet falls away into a pit. But instead of plunging down forever, it opens out again to form a second surface underneath.

This is unexpected. What are we to make of the lower part? What is the significance of the region of space down there? The best way of describing the lower surface is as "another universe," although a proper understanding of these features didn't come until 1960, with the work of George Szekeres in Australia and Martin Kruskal in the United States.

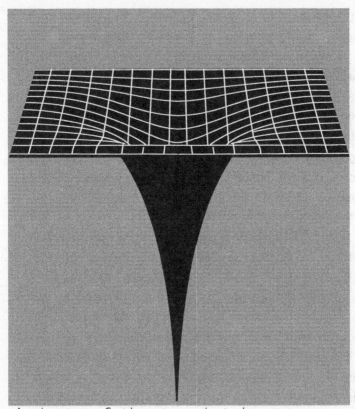

A lot of thought has
been given to what
lies at the bottom of
the pit.

A structure of this
general form is called a
wormhole.

Intriguing though it is, the "other universe" that links to ours via the wormhole shouldn't be taken too seriously, being the product of an idealized mathematical model. It traces back to 1916, and a solution of Einstein's equations discovered by Karl Schwarzschild to represent the gravitational field in the empty space outside a star. It doesn't apply to the interior of the star. If you try patching together Schwarzschild's solution with another describing the star's innards, the whole bottom half of the wormhole is eliminated. Even when you allow the star to collapse to a singularity, you don't create the bottom sheet.

The only way the entire wormhole could make physical sense is if somehow the universe was made like this, with the wormhole embedded in it, coutesy of Mother Nature. Even then, there is a complication, because the wormhole doesn't just sit there: it changes with time. Initially, the two universes are detached. Then they connect together at a single point corresponding to a singularity, where the curvature of space is infinite. This is just like the singularity that results when a ball collapses to a point of infinite density, only in this case there is no collapsing ball, only empty space.

From this singular beginning, the wormhole throat then opens out, but only for a limited duration, after which it closes up again, and the two universes disconnect. Crucially, this sequence happens so fast that nothing can get through the wormhole before the throat pinches off. Not even light can pass from one universe to another, so an observer in our universe wouldn't even be able to see the "other universe," let alone visit it. This makes the existence of the "other universe" rather hypothetical, since the two universes—top and bottom sheets of the wormhole—couldn't affect each other in any way. Any astronaut stupid enough to jump into the black hole would end up hitting the singularity at the center and being totally obliterated.

⟋ Curved Spacetime

I've described what it means for space and time individually to be warped by gravity, but it's more accurate to consider them together in a unified description.

The concept of *spacetime* is not especially hard to envisage, and here time is drawn vertically and space horizontally. For ease of representation, I have shown only one space dimension. The heavy lines on the diagram are the paths in spacetime of physical objects. Path 1 is a body that remains at rest as time goes along. Path 2 is a body that moves at constant speed to the right. Path 3 shows a body accelerating to the right. The wiggly line labeled "Path 4" depicts a body that moves back and forth.

What can we say about the physics underlying these differently shaped paths? Since Newton, it has been known that a body will accelerate only if a force acts on it, so curved paths in a spacetime diagram demand physical

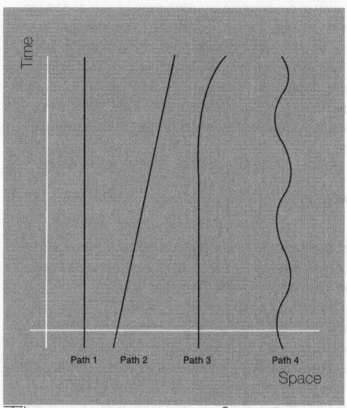

The concept of *space-time* is not especially hard to envisage.

forces. Path 4, for example, implies a push-pull alternating force to make the body zigzag back and forth.

Gravity is one among a number of physical forces. Einstein's great insight was to spot the significance of the fact that gravity differs from other forces in a crucial respect: it affects all bodies equally. There is a famous story about Galileo dropping heavy and light bodies from the Leaning Tower of Pisa to convince skeptics that the objects would hit the ground together. Translated into a spacetime diagram, this means that when gravity is the accelerating force, all bodies (light, heavy, hot, cold, living, dead . . .) will follow the same path. The same would not be true of, for example, an electric field, which would accelerate charged particles but leave uncharged particles to follow straight spacetime paths.

Einstein reasoned that if the effect of gravity on moving bodies is the same for all, it is better to represent the gravitational field not as a force, but as a geometrical property of spacetime. This mysterious-sounding idea is easily understood. Here I have depicted the two-dimensional sheet containing my spacetime diagram, but it is no longer flat—it is curved or warped. It should be clear that distorting the sheet in the manner shown mimics the effect of curving the paths. In other words, the wiggles in a path may be achieved by drawing either a wiggly line on a flat sheet or a "straight" line on a curvy sheet. Straight here means "straightest," that is, the shortest path between two points across a curved sheet. Einstein proposed that

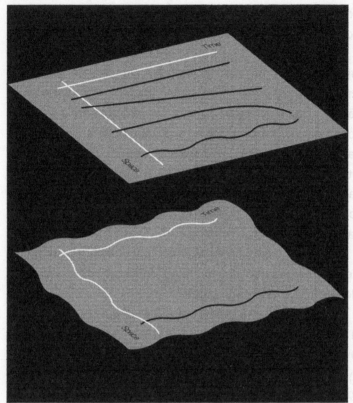

It is better to represent
the gravitational field
not as a force, but as
a geometrical property
of spacetime.

in the case of gravity, it is better to think of the gravitational field geometrically, rather than as a force acting in "flat" spacetime. Of course, to do this properly entails extending the notion of spacetime curvature from the two-dimensional sheet shown to four dimensions (three of space and one of time), but this is straightforward mathematically.

⌀ Wormholes: Portals to Another Universe?

To serve as a time machine, a wormhole has to be traversable: the time traveler has to be able to pass through it and emerge intact. This is impossible with the Schwarzschild wormhole, as it pinches off before anything can get through it. However, that entire model assumed empty space and exact spherical symmetry. Suppose we relax these assumptions.

In the 1960s physicists and mathematicians began studying the properties of *spinning* black holes. These bulge around the waist in the same manner as rotating planets because of centrifugal force. Now centrifugal force acts to oppose gravity. The reason the Schwarzschild wormhole pinches off so fast is the intense gravity within it. With rotation present the pinching effect is diminished, raising the prospect that the wormhole throat might stay open long enough for something—or someone—to get through. Forty years ago it looked as though spinning black holes would provide traversable wormholes, at least from the idealized mathematical models in use at the time. There was much discussion about what might await an astronaut who tumbled into a spinning black hole and emerged in another universe.

On closer inspection, several problems surfaced with this scenario. The first is practical. Any astronaut leaping into a black hole risks being mangled by the intense gravitational forces. To see why, imagine yourself jumping out of an airplane feet first. Because the Earth's gravity diminishes with height, your feet, being closest to the ground, will be pulled down a little more strongly than your head, so your body is stretched slightly lengthwise; at the same time, your shoulders get squeezed together, because each shoulder is pulled toward the center of the Earth, and the Earth's curvature means they try to fall on converging paths. Effectively you are turned into spaghetti.

It was just such stretch-and-squeeze gravitational forces that ripped comet Shoemaker-Levy 9 into fragments before it plunged into Jupiter in 1994. Near a solar mass black hole the effects would be so strong they'd "spaghettify" an astronaut in pretty short order. Spaghettification is less likely if the hole is bigger. You could just about survive falling to the surface of a black hole with ten thousand solar masses. A supermassive black hole a billion kilometers across would be no problem, but such an object would have a mass equal to a small galaxy—not

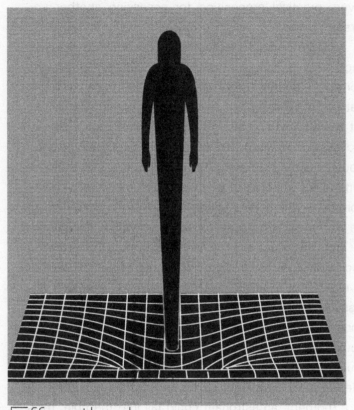

Effectively you are turned into spaghetti.

a very practical proposition for accessing another universe.

A more serious problem with traversing a spinning black hole is that the idealized model containing the wormhole ignores the effects of any matter or radiation that might be around. Not only can the astronaut fall in, but so can anything else that comes along, such as cosmic rays and starlight. The intense gravity of the hole enormously boosts the energy of these as it sucks them in, forming an impenetrable wall across the throat of the wormhole. The gravity of the wall would probably cause the wormhole to collapse, sealing it off with a singularity.

And that's not all. The centrifugal force of a rotating black hole does combat the inward pull of gravity, but not so much that a singularity is prevented. On p. 38, I discussed how an exactly spherical ball of matter would implode to a single point of infinite density. A spinning ball wouldn't be spherical, because of the bulge around the equator. Instead, it collapses to form a ring singularity inside the hole. If we ignore the above-mentioned problems for a moment, an astronaut could fall into the hole, miss the singularity, and come out in another universe.

The idea that an astronaut might inspect a singularity and live to tell the tale strikes horror into the heart of the physicist. Taken at face value, a black hole singularity is an entity with infinite density and infinite space curvature. Space and time cannot be continued through it. Singularities are therefore edges or boundaries to space and/or time. There is literally *nothing* beyond them: they are places where physical objects and influences could leave or enter the universe. A chunk of matter that hits a singularity and ceases to exist is bad enough, but what about a chunk of matter that spontaneously pops out of a singularity?

The idea of having a region of space from which anything at all might emerge without cause and without warning is pretty startling. It would represent nothing less than a breakdown of the rational order of the cosmos. For this reason, Sir Roger Penrose proposed a law of nature to ban such unwelcome invasions. He conjectured that singularities are so obscene they will always be decently clothed by black holes. In that way, nobody in the outside universe

would be able to see a singularity. No uncaused physical influences could emanate into the wider cosmos to wreak havoc. Never would an edge to spacetime be exposed to public gaze.

Penrose called this ban his *cosmic censorship hypothesis*:

Let there be no naked singularities!

And this is where the trouble lies concerning spinning black holes. If you could fall into one, zoom past the ring singularity, and come out again in another universe, then all the gremlins from the singularity could come out with you. The singularity would be naked to the other universe, in defiance of cosmic censorship.

Now it has to be said that this isn't a watertight no-go theorem for traversing spinning black holes. Nobody has proved the cosmic censorship hypothesis; it might be wrong. Also, a singularity could be a mathematical fiction: perhaps the general theory of relativity, or even the concept of spacetime, breaks down before a singularity forms. Still, for all the above reasons, using a spinning black hole as a gateway to another universe looks decidedly suspect. If the aim is to find a safely traversable wormhole, something else is called for—something to combat gravity with more oomph.

⟋ How to Make a Traversable Wormhole

The concept of time travel began with a work of science fiction, and it has remained firmly in the realm of fiction until recently. Curiously, the trigger that transformed time travel into the business of serious science was another work of fiction. In the mid-1980s the astrophysicist Carl Sagan wrote his novel *Contact*, which later became a Hollywood movie starring Jodie Foster. The story is not actually about time travel but concerns a radio message received from an advanced alien civilization. The message contains the design of a machine to create a wormhole in space between Earth and the star Vega, twenty-six light-years away. The wormhole is then used by a team of scientists to travel and meet the aliens. Sagan employed the wormhole idea as a fictional device to get around that old bugbear of sci-fi—the finite speed of light. In *Contact*, the scientists reach Vega in only a few minutes.

Sagan's wormhole differs in one important detail from the ones I discussed above. The black hole wormholes were conjectured to be a gateway to another universe. Sagan's wormhole is a tunnel linking two points in the *same* universe. Sagan gave scant details about how such a wormhole was to be constructed. In the movie version, Jodie Foster climbs aboard a capsule and gets dropped into what looks like a gigantic kitchen mixer, where-

upon she zooms through a narrow tunnel and emerges partway across the galaxy. It looks great, but is it feasible? Intrigued, Sagan wanted to know whether using a wormhole as a shortcut through interstellar space had any scientific credibility, so he approached

his friend the theoretical physicist Kip Thorne at the Cali-
fornia Institute of Technology.

Thorne and his colleagues agreed to check out what
would be needed to make Sagan's vision a reality. They
did this by adopting a sort of reverse engineering approach to gravitational theory. Normally a physicist consid-
ers a gravitating object—a star, say—and uses the general
theory of relativity to work out the gravitational field it gen-
erates, which in turn predicts how the space near the object will curve.

For this project, Thorne started by writing down the
answer first. He knew the sort of geometry of space
needed: something shaped like a wormhole with two
spherical mouths. But it had to be a *benign* wormhole—
one that stayed open long enough to allow Jodie Foster to
get through, and did not rip her apart with gravitational
forces or incinerate her with surfaces of infinite energy.
Obviously the sort of wormholes I discussed above
wouldn't do. Then Thorne asked what type of matter
would be needed to generate this benign wormhole.

It soon became clear that any familiar form of matter (water, diamond, hydrogen gas, light, neutrinos . . .)
was out of the question. In all cases it would make the
throat of the wormhole collapse before anything could
traverse it. Clearly, some exotic form of matter would be
needed.

It's not hard to figure out what. If a wormhole is traversable, it must have an exit as well as an entrance. In that

case, it should be possible to shine light through it. The reason a black hole has no way out is that its gravity bends light inward, trapping it and focusing it down onto a singularity. As the wormhole allows light to come out the other end, somewhere inside it light would have to be *defocused*—that is, bent outward.

Thorne realized that the way to do this was to use some sort of *antigravity*. This is no surprise. Something powerful is needed to shore up the wormhole, to combat gravity's inexorable tendency to crush the wormhole and pinch it off in a singularity. *Antigravitating* matter is one answer. But does it exist?

Well, it has long existed in folklore. Levitation is an ancient myth and is featured in many world religions and mystical beliefs. Antigravity remains a favorite idea among UFO buffs for alien spacecraft propulsion. It also attracts a variety of independent thinkers, wacky inventors, and visionary venture capitalists, fixated by the dream of nullifying the Earth's gravity and floating to the stars without the need for rockets. Antigravity also crops up in science fiction: H. G. Wells envisaged a sort of gravity shield (called Cavorite) in *The First Men in the Moon.*

The first appearance of antigravity in science was provided by Einstein. In 1917, he doctored his own general theory of relativity to incorporate a repulsive form of gravitation. He did this to produce a model of the universe. At that time, nobody knew the universe was expanding. Einstein was puzzled (as was Newton) about how the universe could be static when the only truly cosmic force is gravitation, which is universally attractive. So he added an extra term to his gravitational field equations to describe a type of antigravity. By balancing the attractive force of normal gravity with the repulsive force of antigravity, a static universe might result.

Once Einstein discovered that the universe was not static but expanding, he abandoned the repulsive force, calling it the biggest blunder of his life. Ironically, he could have been right after all. Although antigravity may not be needed for a static universe any more, the force may yet exist, and recent astronomical evidence suggests that, in fact, it does. However, in its pervasive cosmic form, Einstein's antigravity is far too feeble to help make a traversable wormhole.

Antigravity crops up in other branches of physics too, though only under unusual circumstances. The basic idea is easy to grasp. In normal matter, mass is the source of its gravity. Because of the link between mass and energy ($E = mc^2$) all forms of energy gravitate. If what you want is antigravity, this can be produced by *negative* energy:

Positive energy gravitates.
Negative energy antigravitates.

At first sight, negative energy seems as mysterious as a negative lunch. Surely, either you have some lunch or you don't. How can you have less than no lunch?

The answer lies with the definition of zero energy. Because energy gravitates, zero energy must correspond to a state with no gravitational field whatsoever. In the general theory of relativity, that condition implies no space-warps and no timewarps—spacetime is precisely flat. So if we can engineer a physical situation in which the energy is *less* than such a zero energy state, the energy will be negative, and the state will antigravitate.

Imagine a box made of ordinary matter filled with enough negative energy to make the total mass-energy negative. Would it then fly upward instead of falling downward? Unfortunately not. True, the box would feel an upward gravitational force, but because its mass is negative, it would actually move in the opposite direction, downward! So negative energy falls just like positive energy. It can't be used to soar to the stars.

However, the gravity field that the negative energy itself creates is certainly repulsive. A ball made of normal matter placed near the box would be accelerated *away* from it. If the Earth were made of negative energy, we should all be shot into space.

In chapter 3 I shall explain how negative-energy states can be created, but for now let's assume that some suitable exotic matter is available, and is stuffed into the throat of the wormhole. If it antigravitates powerfully enough it will stop the throat from collapsing and allow light and perhaps even astronauts to pass through. The final form of the wormhole can then be represented as a flexible two-dimensional sheet, but this time the sheet is bent right around until the two ends come close together and then get connected through the wormhole. In this manner, points A and B that lie far apart in space—perhaps many light years—can be joined by a short wormhole, exactly as in *Contact*.

Bending the sheet around in the manner shown seems like drastically curving a large portion of the universe until it is almost folded back on itself, a task that would tax even a supercivilization. In fact, the representation is misleading in this respect. It is true that gravity curves space, but the folding-over curvature here is *not* a gravitational spacewarp. The act of folding the sheet, or even rolling it into a cylinder, does not affect the geometrical properties within the sheet itself.

To see this, imagine drawing geometrical figures on the sheet—triangles and squares, say. When the sheet is simply folded over, there is no stretching or shrinking. Nothing changes within the surface: all the angles remain

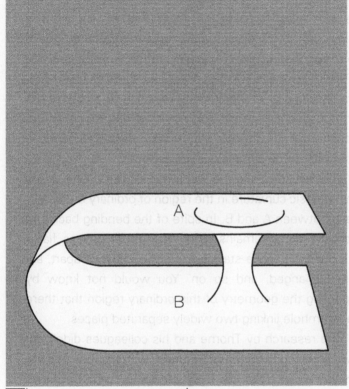

The two ends come close together and then get connected through the wormhole.

the same, squares stay squares, and so forth. Contrast this with trying to paste the sheet onto the surface of a sphere. In that case you would have to stretch or crease the sheet, thereby altering angles, deforming squares, and so on. (The reverse is also true; just think of the distortions involved with Mercator's projection of a map of the Earth.) Cylindrical surfaces do not have intrinsic curvature, but spherical surfaces do. Similar statements can be made about their three-dimensional equivalents.

When it comes to the wormhole depicted here, there is no intrinsic curvature in the region of ordinary or "outer" space between A and B. In spite of the bending back, the geometry there remains as it was—more or less flat—with points in space staying the same distance apart, angles unchanged, and so on. You would not know by inspecting the geometry of this ordinary region that there is a wormhole linking two widely separated places.

The research by Thorne and his colleagues didn't uncover anything fundamentally wrong with the idea of a traversable wormhole, so long as some form of exotic matter could be deployed. And the matter needed was not *too* exotic. Some known physical systems are believed actually to possess it, albeit in tiny amounts. This was a very significant discovery. It did not prove that traversable wormholes definitely could exist, but it didn't rule them out either.

That was exciting enough already. More was to follow, though. Once the researchers digested the possibility of a wormhole in space, it dawned on them that if one were somehow made, it could also serve as a time machine. As with a black hole, the gravitational field of a wormhole can act as a means to reach the future. However, the wormhole can do more: it can also be used to travel back to the past. By passing through the wormhole from A to B, it is possible to go backward in time. And returning rapidly across ordinary space, you could get back to A before you left. At last physicists had found a plausible way to travel both back and forth in time. But how might a wormhole time machine be made?

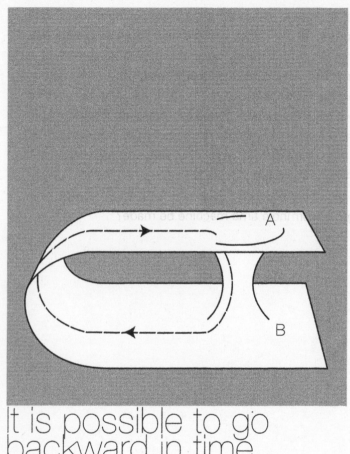

It is possible to go
backward in time.

3. How to Build a Time Machine

Some scientists think wormholes may have formed naturally in the big bang, so that an advanced spacefaring civilization might expect to discover one in the galaxy and commandeer it for the purpose of making a time machine. However, it would obviously be more convenient to manufacture one artificially. But how do you go about constructing a traversable wormhole and turning it into a time machine? Here is a possible scheme for a production factory. The construction is a four-stage process, represented by four workshops, containing, respectively, a collider, an imploder, an inflator, and a differentiator. Let me explain the purpose of each in turn.

Wormholes and time machines today are regarded as outrageous by most physicists. Kip Thorne

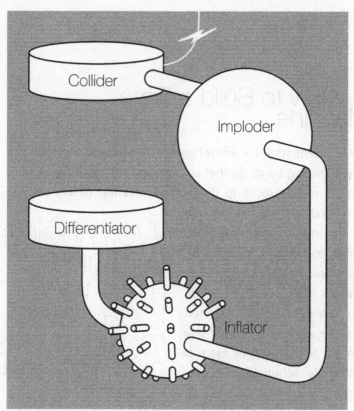

How do you go about constructing a traversable wormhole and turning it into a time machine?

$\mathcal{O}$ The Collider

An obvious and rather fundamental problem stands in the way of any attempt to make a wormhole in a normal region of spacetime. Think how you would go about producing one using a sheet of paper. Although the sheet may be folded over until it touches itself, there is no way you can join the two surfaces to form the wormhole without cutting the paper and pasting it together again. No matter how much you twist and turn, pull and push, at some stage an incision has to be made. The problem is the same as that of taking a ball of putty and turning it into a doughnut shape. You can't avoid tearing at the putty to make a hole. This difficulty is independent of the precise geometry involved; it is a property of the *topology* of the system.

In the case of a wormhole in space, the "sheet" is space itself. It is important to realize that the wormhole isn't a hole through anything—it is actually made of space.

So how do you carry out surgery on space? Nobody knows how to do this on a large scale. Think what it would mean to cut into the space near Earth. Before you pasted it together again, you would have exposed a raw edge. As we saw in chapter 2, edges to space are possible; they are called singularities, and are seriously bad news. That was the way the Schwarzschild wormhole was created— at a singularity with infinite density, but it was buried inside a black hole. An *exposed* edge of the sort needed to

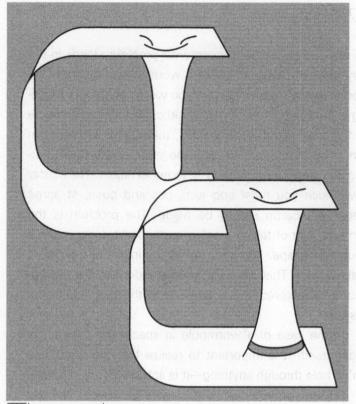

There is no way
you can join the two
surfaces . . . without
cutting the paper and
pasting it together
again.

make a traversable wormhole would be a naked singularity; such a thing might play havoc with nature. In any case, building a wormhole by making spacetime singularities is too violent; we want to do the job in a more controlled manner.

A better method is to employ the quantum vacuum. Quantum mechanics is based on Heisenberg's uncertainty principle, which predicts that all physical quantities fluctuate randomly. On an atomic scale, properties like the speed or energy of a particle can be highly uncertain. As a general rule, the smaller the scale, the bigger the fluctuations. At some tiny size, quantum uncertainty will become so big that it will produce significant gravitational effects. We can see that by considering energy. According to Heisenberg's uncertainty principle, energy will be uncertain for brief durations, which means that its value can change unpredictably. One way to think of this is in terms of borrowing. An electron, say, can borrow energy from nature, so long as it pays it back in a short while. The essence of the uncertainty principle is that the bigger the loan, the quicker the payback time.

Putting in the numbers, you find that for a duration as brief as a ten-million-trillion-trillion-trillionth of a second, known as the Planck time (after the German physicist Max Planck, who founded quantum theory), so much energy can be borrowed that its mass will seriously warp spacetime, sculpting it into elaborate structures. It's not clear exactly what the upshot is, but John Wheeler has painted a vivid picture of a labyrinth of tubes and tunnels, which he evocatively calls *spacetime foam*. The size of these structures is about a billion-trillion-trillionth of a centimeter (known as the Planck length), which is exceedingly small—twenty powers of ten smaller than an atomic nucleus, in fact.

Now the quantum wormholes I'm talking about aren't permanently embedded in space, because they live on borrowed time. The energy needed to bend space into a complex foam is on loan via the Heisenberg uncertainty principle. So quantum wormholes don't last; they come and go at a frenetic pace.

But what about the cutting-and-pasting problem mentioned above? In the quantum domain, the difficulty of singularities is circumvented. Changes in topology become submerged in the overall quantum fuzziness of everything. Trying to pin down where space might be ripped open is as pointless as attempting to locate where an electron is in an atomic orbit. Such things are intrinsically indeterminate in quantum physics.

To distinguish the temporary, ghostly quantum wormholes from big, permanent, real ones, physicists refer to the former as *virtual*. A virtual wormhole is one that exists fleetingly, courtesy of the Heisenberg uncertainty principle. Kip Thorne has suggested that an advanced civilization might develop the technology to reach into the spacetime foam, pluck out a virtual wormhole, and expand it into a big, permanent one. This means gaining control over nature on a scale of size about fifteen powers of ten smaller than our current capabilities.

A direct approach looks hopeless. However, there may be a way to do it indirectly. One problem about harvesting virtual wormholes from the spacetime foam is that they typically last for only a Planck time before vanishing. To create a permanent wormhole, we must artificially inject enough energy into the spacetime foam to "clear" the loan on the virtual wormhole's behalf, and thereby promote it to a real one. It sounds fanciful, but we do a similar thing all the time in radio transmitters. An electric field can be envisaged as a cloud of virtual photons scurrying around a charged particle such as an electron. If energy is fed into the system—say, by accelerating the electron in a wire—then some of the virtual photons are turned into real photons and flow away from the wire in the form of radio waves.

The Heisenberg uncertainty principle also has important implications for the nature of empty space: it means there is no such thing as a perfect vacuum. Even when you have removed all particles of matter and all photons, there will still be virtual photons (and virtual versions of all other types of particles) popping into temporary existence. Virtual particles permeate all of space, filling it with a seething ferment of quantum activity. What may appear at first to be total emptiness is, in fact, a beehive of fluctuating ghosts, appearing and disappearing in an unpredictable frolic. And this isn't just theory. Virtual photons manifest themselves physically in a number of ways. For example, they jostle electrons in atomic orbits, producing small but measurable changes in the energy levels. They also produce the so-called Casimir effect, which I shall discuss on pp. 80–82.

The collider is the first step toward delivering the required energy to the spacetime foam. It involves a heavy-ion accelerator of the type employed at the Brookhaven National Laboratory on Long Island, New York. This machine is designed to boost the nuclei of atoms such as gold and uranium to colossal energies, and then collide them head-on. The nuclei are confined by magnetic fields to a ring-shaped vacuum tube, in which they are accelerated using electric pulses, and arranged so that counterrotating beams of nuclei are brought into high-speed contact. The collisions are designed to be so violent that they briefly re-create the conditions that prevailed in the universe about a microsecond after the big bang, when the temperature was a searing ten trillion degrees.

When the nuclei smash into each other, their constituent protons and neutrons are pulverized, creating a bubble of energized fragments known as a quark-gluon plasma. (Sometimes the dramatic phrase "melting the quantum vacuum" is used.) In effect, the components of the nuclear particles—quarks and gluons—part company and mill around inside an amorphous blob.

The quark-gluon plasma bubble having been created, the next step is to pass it on to the imploder.

⟨ The Imploder

Although by human standards a quark-gluon plasma is highly energized, it is a long way from our present requirements. The enormous temperature of ten trillion degrees inside the bubble is still about nineteen powers of ten too low to affect the spacetime foam. To boost the temperature up to Planck values we need to compress the bubble by a factor of a billion billion. Surprisingly, the total energy involved in achieving this is very modest—about ten billion joules, equivalent to the total output of a typical power station for only a few seconds. So energy is not a limiting factor at this step. The challenge is to concentrate that much energy in such a small object.

It's not clear how this might be done, but explosive magnetic pinching could offer a way. Magnetic fields are used to confine conventional low-energy plasmas, such as ionized gases. If the field is intensified, the plasma gets squeezed. Scientists began experimenting with this technique, known as the Z-pinch, in the early 1950s, as part of the program to develop controlled nuclear fusion. An intense electric current is passed through deuterium gas in a chamber, rapidly ionizing it. The magnetic field of the current then violently pinches the resulting plasma and heats it to millions of degrees. The most refined Z-pinch system in current use is at the Sandia National Laboratories in New Mexico, where electrical pulses of fifty trillion watts from a bank of charged capacitors are concentrated onto ultrathin tungsten wires.

The sort of compression needed to reach Planck temperatures demands far more punch than the Sandia system. A collection of thermonuclear bombs arranged in a spherical pattern centered on its target might be able to concentrate a magnetic field enough to implode the quark-gluon bubble. To repeat, the total energy required is not great; it merely needs to be directed at the target bubble rather than spewing into the surroundings. Assuming the concentration problem can be solved, the net effect will be to create a tiny ball with a density of about ten trillion trillion trillion trillion trillion trillion trillion trillion kilograms per cubic meter, or some eighty powers of ten greater than the density of nuclear matter. This is enough to rival the vast energy fluctuations permitted at the Planck length: a billion-trillion-trillionth of a centimeter—the distance light will travel in a Planck time. The result, it is hoped, would be to form either a minute black hole or a wormhole that would become the seed for growing the time mchine.

To implement the compression, some serious problems of basic physics need to be addressed alongside the engineering challenges. Quantum field theory suggests that if a magnetic field gets too strong it may start to create subatomic particles and thereby dissipate itself. Also, magnetic pinching is notoriously unstable. These difficulties could possibly be circumvented by using another type of field altogether, such as the so-called Higgs field that is being sought eagerly by particle physicists.

Alternatively, an accelerator might be employed in place of an imploder. With conventional electromagnetic technology, Planck energies could be attained only by building an accelerator as big as the solar system, but radical new accelerator techniques might achieve very high energies with much more compact equipment. Also, some theories suggest that major changes to space and time might manifest themselves at energies much less than the Planck energy, and may even lie within the reach of foreseeable technology. If gravity can be manipulated at modest energies, wormholes may form without the need for such enormous compression or acceleration.

Once a real, albeit minuscule, wormhole is produced, the next step is to inflate it to manageable dimensions.

⏝ The Inflator

Since a Planck-sized wormhole is practically useless, some method has to be used to drastically enlarge it. As we have seen, the crucial ingredient in stabilizing a traversable wormhole is some sort of exotic matter with anti-gravitational properties. So the next step in the production process is to feed the nascent microscopic wormhole with exotic matter. Its antigravity would then push the throat of the wormhole outward, enlarging the size. The scheme I am proposing uses a bank of high-powered lasers with an ultrafast rotating mirror system.

Before writing any more about the inflator, I need to explain a bit more about antigravity. In chapter 2 I discussed how it can be produced by negative energy (see p. 63). So how do you create negative energy? A simple method was discovered by the Dutch physicist Hendrik Casimir in 1948. This is what you do. Take two sheets of metal and place them close to each other, face-to-face. Secure them so they don't move. Now enclose the entire system in a large, thick metal box from which all other material (including gases and electrically charged and neutral particles) has been removed, and cool to absolute zero (-273 degrees Celsius). The slab of empty space between the plates now contains negative energy.

THE EXPLANATION

The Casimir effect is a phenomenon of the quantum vacuum. Strictly, I should not cite it as an example of exotic *matter*, because it refers to a state of empty space. But this is a terminological quibble: the distinctions between field excitations, matter, and emptiness are very blurred in quantum physics.

This is why the negative Casimir energy arises. The apparently empty space between the plates is not a complete vacuum, but populated by a seething mass of virtual photons. Like their real counterparts, virtual photons re-

bound from the metal plates. Being sandwiched between the plates, they are not free to move in any direction, so this restriction affects the variety of virtual photons that can inhabit the interplate region compared with the space outside the two plates. In effect, the possibility for some sorts of virtual photons to exist is eliminated by the presence of the plates. As a result, the total energy being "borrowed" (via Heisenberg's uncertainty principle) in the interplate region is a bit less than it would have been without the plates. If we agree that apparently empty space without any plates has exactly zero energy, then the region between the plates must have negative energy. The negative energy manifests itself by producing a tiny force of attraction between the plates.

CAN IT BE DONE?

Yes! It has been done. The Casimir force was first measured in the laboratory in 1958 and has been studied many times since. In these experiments no attempt is made to remove all the (much larger and positive) other sources of energy pervading the system, since the purpose of the experiments is to test Casimir's prediction, not actually to create a region of negative energy. The experiments confirm the theory. The force of attraction between two perfectly reflecting plates one meter square held a hundreth of a millimeter apart is equivalent to the weight of just a millionth of a gram. But the force grows larger as the plates get closer. With real metal sheets, which are never perfectly flat, the effect is complicated by other factors long before the Casimir force becomes large. However, this hasn't stopped some imaginative theorists from toying with ideas of using the Casimir effect, and other quantum vacuum effects, as the basis of a spacecraft propulsion system.

OTHER WAYS OF MAKING NEGATIVE ENERGY

The Casimir effect is the most famous, and easiest, method of producing negative energy by disturbing the quantum vacuum. But there are other ways too. Negative quantum vacuum energy can be created by a single reflecting surface (that is, a mirror) if it is moved vigorously. In the mid-1970s I studied this "moving mirror" effect in great detail with my collaborator Stephen Fulling. We restricted our work to a simple one-dimensional model, but similar results are likely to apply in real three-dimensional space too. We found that if a mirror moves with increasing acceleration, a flux of negative energy emanates from its surface and flows out into the space ahead of the mir-

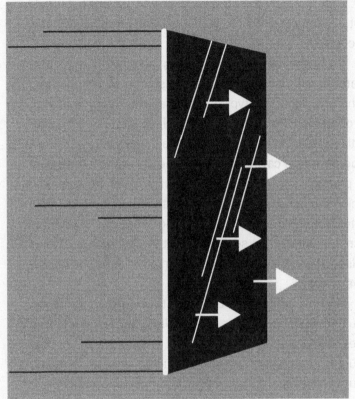

Negative quantum
vacuum energy can be
created by a single
reflecting surface.

ror. Unfortunately the effect is exceedingly small; it would not be a practical way to generate large amounts of negative energy.

Probably the most promising negative-energy generator is the laser—a high-energy source of very pure, or coherent, light. In a typical setup, a laser beam is passed through a crystal of lithium niobate, shaped like a cylinder with rounded silvered ends to reflect light, thus forming a type of optical cavity resonator. The crystal has the effect of producing a secondary beam at a lower frequency in which the pattern of photons is rearranged into pairs. This is technically known as "squeezing" the light. Viewed in terms of energy, the squeezed light that emerges contains pulses of negative energy interspersed with pulses of positive energy.

Crystals are not the only way to squeeze light. If you could manufacture very reliable light pulses containing specifically one, two, three, . . . photons apiece, they could be combined together in such a way as to create squeezed states to order. By superimposing many such states, bursts of intense negative energy could, in theory, be produced.

The main snag about using lasers is the short duration of the negative-energy pulses. Typically, one might last for 10^{-15} second, after which it is followed by a positive-energy pulse of similar duration. Some method has to be found to separate the positive from the negative parts of the laser beam. The inflator facility I am proposing uses a set of rapidly rotating mirrors with the light striking each mirror surface at a very shallow angle. The rotation would ensure that the negative-energy portion of the beam was reflected at a slightly different angle from the positive-energy part. Far away from the mirror there would be a small separation of the positive and negative components of the beam, and a further system of reflectors would allow only the negative part to be directed into the wormhole.

With current laser technology, the numbers look disappointing. Even if negative energy can be directed at a wormhole in a sustained manner, and somehow trapped within the throat, it would require a vast length of time to accumulate enough of it to make a macroscopic wormhole. The theoretical physicist Matt Visser has estimated that a one-meter-diameter wormhole would need a negative energy equivalent to the mass of Jupiter. With perfect separation of positive and negative energy from a million terawatt lasers running flat out and continuously, it would still take far longer than the age of the universe to build up that much negative energy.

◯ Other Inflator Devices

Negative quantum vacuum energy also occurs as a by-product of certain gravitational fields. A simple example is the gravitational field of the Earth, which produces a cloud of negative energy around it by dragging some of the virtual photons downward. In the case of the Earth, the effect is extremely small. But as the gravitational field rises, so the negative-energy cloud grows in strength. Near the surface of a black hole it is enormous. Because there is no material surface to the hole, only empty space, this negative energy flows into the black hole in a steady stream. In effect, the black hole vacuums up the quantum vacuum!

A solar-mass black hole with a radius of three kilometers sucks up negative energy at a rate of a billion-billion-billionth of a joule per second. Still pretty feeble. But the smaller the black hole, the stronger the gravity at its surface, and the more intense the negative energy that surrounds it. A black hole the size of an atomic nucleus (and the mass of a mountain) would swallow negative energy at about a billion joules per second, creating a million-kilowatt energy sink.

The existence of negative energy near black holes was guessed by Stephen Hawking in 1974. Hawking predicted that a black hole should glow faintly with heat radiation. The radiated energy has to come from somewhere, and since nothing (even energy) can get out of a black hole, it seemed the only explanation must be that negative energy flows *into* it. The following year, William Unruh, Stephen Fulling, and I confirmed this prediction by computing the energy near a black hole in a simplified two-dimensional mathematical model. We found that there is, indeed, a negative-energy flux into the hole at a

rate that exactly compensates for the heat radiation coming off it.

Stationing our time machine factory near the surface of a black hole to soak up the negative energy is hardly feasible, but the very existence of negative quantum energy generated by gravitational fields is highly significant. Since the wormhole itself will have a strong gravitational field, it may be that it will generate the required negative energy from the quantum vacuum of its own accord. Nobody yet knows whether this is possible or not. If it is, the wormhole might be induced to self-inflate, with very little exotic matter input. The laser system could be used to start the process, configuring the geometry of the microscopic wormhole to the appropriate shape, and after that nature would do the rest, delivering a large wormhole for free. The wormhole would loom out of the spacetime foam as, so to speak, a "free lunch." If that seems surprising, remember that negative energy has negative mass, so that the total wormhole mass might be close to zero. In other words, there may be little or no overall energy cost in producing a wormhole: the negative-energy parts pay for the positive-energy parts. In that case, a large wormhole could generate itself spontaneously, with a little bit of fine-tuning from the inflator factory engineers.

The primary requirement in making a serviceable time machine is keeping the wormhole throat open. But to use a wormhole as an effective transportation device demands that it be more than merely a gateway to other times and places. A human being has to be able to squeeze through and come out smiling. To avoid spaghettifying its users, a wormhole's gravitational field needs to be gentle. Also, the duration of the ride should be reasonable. A journey into the past that takes the time traveler one hundred years to complete would not be very attractive. So long wormholes are out.

These two additional requirements impose further restrictions on the exotic matter in the wormhole. Experts quarrel over just how exotic the exotic matter must be, whether it should be confined to a small region deep in the throat of the wormhole or allowed to emanate from the mouths, how it can be so confined, whether radiation issuing from one mouth could enter the other and endlessly cycle back on itself, and a host of other technicalities.

Assuming that all these difficulties can be overcome, and we have a safe, short wormhole in the storage bay of the inflator factory, the final step is to convert it into a time machine. That is the job of the differentiator.

$\mathcal{O}$ The Differentiator

To turn a wormhole into a time machine you have to establish a permanent time difference between the two ends. The simplest technique is to use the ordinary time dilation—or twins—effect. To do this, the wormhole is given an electric charge (for example, by firing electrons into it) when it is still quite small—say, the size of a subatomic particle. One mouth of the wormhole is then fed into an ordinary circular particle accelerator and whirled around at very near the speed of light, while the other is held still. This produces a growing temporal discrepancy between the two mouths of the wormhole. The process is allowed to continue for, say, ten years, at which time the moving mouth is brought to rest and allowed to approach the other wormhole mouth. The wormhole is now able to send particles of matter back in time for up to ten years. In the final step of the process, the wormhole is returned to the inflator factory to be expanded to a size large enough for a human being to traverse—say, ten meters in diameter. Meanwhile, the length of the wormhole is kept as short as possible.

Another way to turn the wormhole into a time machine is to use the gravitational field of a neutron star rather than an accelerator as a differentiator. This is how it works. Imagine a rather short wormhole, let's say ten meters in length. Tow one mouth—call it A—to the close vicinity of a neutron star a few light years away, and leave the other end, B, parked in our solar system. Keep them in place until the gravity timewarp of the neutron star accumulates to the required amount, then tow A back to our solar system and park it next to B. The time machine is now ready for use.

To see why this procedure works, imagine identical clocks placed at each mouth of the wormhole. The gravity of the neutron star stretches time at mouth A; the clock there will run slow. What about the clock at B? Since it is some light-years from the star, its rate should not be affected by the star's gravity, so it should tick considerably faster than the clock at A. But there's a catch. Suppose we look *through the wormhole* from mouth A, located near the star. We then see clock B just a few meters away. So by one route the clock at B is very far from the neutron star; by another it is very near. If it is regarded as very near, then time at B should be slowed by the star's gravity too. There should be very little difference between the clocks' rates at A and B. So which view is right? The answer is, both. Time is, after all, relative, and the situation here is that, viewed through the wormhole, time is about the same at both ends; but, viewed across the "outer" space, the time difference between A and B (clock B is ahead of A) is substantial. If you now jump through the wormhole from A to B, you will jump back 10 years into the past. By returning to A through "normal" space, you could get back to your starting point before you leave. So, once again, by executing a closed loop in space, you also perform a closed loop in time. This time machine is two-way. By going through the wormhole in the other direction—from B to A—you can jump ten years into the future.

Wormhole time travel differs in two crucial respects from the version H. G. Wells presents. First, in *The Time Machine*, the intrepid time traveler throws a lever, and in effect fast-forwards the universe, like a video player, accelerating the "cosmic movie" relative to his mental time. When he gets where he wants to go, he simply hits the stop button. He travels back in time by fast rewinding. The time machine shares in the temporal transport, going back and forth in time with the driver. That is quite unlike the wormhole time machine, which doesn't itself move through time; it is simply part of the cosmic architecture.

Second, in Wells's story, the time traveler doesn't go anywhere in space. But a moment's thought exposes the ambiguity of this arrangement, for in the time spans over

which he travels, the Earth will have moved many light-years across the galaxy. And the galaxy moves relative to others. Since there is no absolute frame of rest against which to measure these movements, the whereabouts of the time machine in space

after the temporal gymnastics is completely indeterminate. The wormhole time machine operates quite differently. Rather than inducing time to run backward, the time
traveler embarks on a journey into space that ends in the
past.

4. How to Make Sense of It All

As nobody has yet produced a knockdown argument to show that time travel is impossible—however daunting the practical difficulties of constructing a time machine may be—the consequences of backward time travel need to be confronted. Writers of science fiction are familiar with the outlandish, even paradoxical consequences that can ensue if people are able to visit the past. So can two-way time travel be incorporated into real science?

> Time present
> and time past
> Are both per-
> haps present
> in time future
> And time future
> contained in
> time past
> T. S. Eliot

⟋ How to Avoid Time Tourists

A much-voiced objection to travel backward in time is that we don't encounter anybody from the future. If it were possible to visit the past, we might expect that our descendants, perhaps thousands of years from now, would build a time machine and come back to observe us, or even to tell us about themselves. Key historical events such as the Crucifixion would have been crowded by throngs of eager witnesses. Discounting reports of ghosts, UFOs, and the like, the apparent absence of time tourists is something of a problem for time travel enthusiasts.

Fortunately this objection is easily met in the case of wormhole time machines. Although wormholes could be used to go back and forth in time, it is not possible to use one to visit a time before the wormhole was constructed. If we built one now, and established, say, a one-hundred-year time difference between the two ends, then in one hundred years someone could revisit 2001. But you couldn't use the wormhole to go back and see the dinosaurs. Only if wormhole time machines already exist in nature—or were made long ago by an alien civilization—could we visit epochs before the present. So if the first wormhole time machine were built in the year 3000, there could not be any time tourists in the year 2000.

⊘ Time Paradoxes

CHANGING THE PAST

Perhaps the most famous of all the time travel paradoxes is the one in which the time traveler goes back in time and murders one of his ancestors, for example, his mother. The problem is then obvious. If his mother dies before giving birth, then the time traveler would never have existed. But in that case he would not be able to carry out the murder. So if the woman lives, she dies, but if she dies, she lives! Either way, contradictory nonsense results.

Many time travel stories have come up against this obvious and thorny problem. In the movie *Back to the Future*, the time traveler, Marty McFly, does not attempt to kill his mother as a young woman. Rather, he becomes embroiled in her love life and risks interfering with her marriage plans, as a result of which he teeters on the edge of obliteration. Of course, his disappearance would not resolve the paradox either, because he would then not have been able to visit the past to interfere in history.

Paradoxes like this arise because the past is causally linked to the present. You cannot change the past without also changing the present; this creates a causal loop. Because the behavior of many physical systems is very sensitive to small changes, even modest tinkering with the past could lead to wholesale changes in the present. Imagine how different the world would be if Adolf Hitler had been assassinated in 1939, or if the tiny genetic mutation that produced the first human had never taken place because the hominid concerned was persuaded to move one centimeter to the left, thus avoiding the crucial cosmic ray that was destined to bring about a species transformation. In Ray Bradbury's story "A Sound of Thunder," a time traveler journeying back to see the dinosaurs kills a single butterfly, and thereby sets in train a series of events that transforms the entire course of history.

Causal loops are not intrinsically paradoxical, so long as they are consistent. *Changing* the past is obviously paradoxical; the past is, after all, past. But *affecting* the past is logically unobjectionable, by which I mean there is no logical impediment to some events being caused by later events, or by a mixture of later and earlier events. For example, imagine a rich venture capitalist whose vast inherited wealth derives from a mysterious benefactor who befriended his great-grandmother a century before. He finances a time machine project and then uses the prototype machine to go back and discover the source of his wealth. He can't resist proving his time-travel credentials by taking a newspaper with him, which he duly presents to his young great-grandmother. Being an enterprising soul, the lady scans the newspaper's stock prices and, with the help of her foreknowledge, makes some shrewd investments. These investments are, of course, the source of her, and her great-grandson's, immense fortune, and the time traveler himself is the mysterious benefactor. No paradox ensues here because the causal loop is self-consistent, and everything fits together neatly.

Paradox looms only when we combine causal loops with unfettered free will. But if the time traveler finds he is simply unable or unwilling to carry out the mischievous deeds that produce inconsistent historical sequences—like murdering his mother—then this particular paradox is avoided.

Why should free will be limited? It may be that you can visit the past, but when you arrive you find yourself continually stymied in what you try to do. If you attempt to kill your mother, perhaps the gun will jam, or you will be arrested first for suspicious behavior, and so forth. Or maybe the wishes that determine your acts during the visit to the past are simply shaped by what is consistent with the future world from which you have come. In any case free will is a slippery concept, hard to reconcile with the laws of physics even without time travel. Many scientists and philosophers dismiss it as an illusion.

It's not necessary for a human being to travel back in time for paradoxical consequences to be triggered. In principle, just a single particle (or any minute physical influence) sent into the past can unleash mayhem. Suppose a sensitive device is programmed to explode if, and only if, it receives a signal from one hour in the future—say, the arrival of a photon with a particular frequency. The device is placed next to the photon emitter. Then sending one such photon back in time would trigger the device and destroy the emitter. But if the emitter is destroyed, the photon is never sent. Again, we get inconsistent accounts.

Even if it is impracticable to build a time machine that could convey humans into the past, it may still be possible to send signals back in time. An early speculation of this sort is based on hypothetical particles called *tachyons*, which can travel faster than light. It is often stated that nothing can go faster than light, but this isn't strictly true. As I explained in chapter 1 (see p. 11), the theory of relativity introduces a light barrier that can't be crossed. A particle of ordinary matter can never be accelerated to a speed faster than light: if you try to do it, the particle just gets heavier and heavier, rather than faster and faster. But the light barrier operates both ways: if something goes faster than light, it can never be slowed below the speed of light. A tachyon is just such an entity stuck on the far side of the light barrier, obliged to travel always at superluminal speed.

If tachyons exist, and can be manipulated, they could be used to send a signal into the past. To do this you would need the help of an accomplice. First you send a signal to a friend using a tachyon beam traveling at, say, ten times the speed of light relative to you. Then the friend immediately sends the signal back again at ten times the speed of light relative to her. If the friend is moving toward you at a large fraction of the speed of light, the return signal will reach you *before* the outgoing one is sent.

What are the prospects that tachyons really exist? Most physicists are extremely skeptical about them. Quite apart from the lack of experimental evidence, they would have some peculiar properties. For example, they would possess imaginary mass (in the mathematical sense), making them hard to reconcile with quantum mechanics. There is no guarantee that they would interact with ordinary matter, in which case it would be impossible to use them to send signals anyway.

Even if tachyons do not exist, wormholes or other devices might be used to send particles back in time. You can then imagine a billiard ball version of the mother paradox. Kip Thorne and his colleagues studied the idea of time-loop billiards. In the modified game, the pockets of the billiard table represent the entrance and exit of a wormhole time machine. Imagine a ball heading for an end pocket, going down, and emerging a few moments earlier from a side pocket in such a way that the ball collides with its earlier self. The collision will then deflect the ball from its initial path and prevent it from entering the end pocket. There is no free will to complicate matters, but, just as in the "matricide" paradox, the described sequence is inconsistent and so won't happen.

But the paradox can be resolved. We can also imagine a slightly different story. Here the ball starts out moving in such a way that it would just miss the end pocket; however, it suffers a glancing blow from a ball emerging from a side pocket. The collision serves to deflect the ball so that it now enters the end pocket and emerges from the side pocket a short while earlier in the guise of the ball that delivers the glancing blow. Thorne showed that this sequence—a ball colliding with its earlier self in a manner designed to create a self-consistent causal loop—is perfectly consistent with the laws of physics. Disturbingly, though, he also showed that there is more than one self-consistent sequence of events. When causal loops are present, the laws of Newtonian mechanics no longer predict a unique reality.

HOW TO MAKE MONEY

Travel into the past takes on an air of absurdity when the time traveler meets his younger self, for then there will be two of him. Note that you would not be surprised to meet your younger self this way, because you would already remember the encounter from your youth.

The age difference need not be great. In principle it could be, say, a day. In this case there would be two virtually identical copies of you. That would be very weird. And it needn't stop there. You could invite your (slightly) younger self to accompany you on a similar trip back another day, when there will be three of you. Nothing prevents this process from being repeated again and again. By making successive hops back in time, the time traveler could accumulate many copies of himself in one place.

Such a scenario suggests a get-rich-quick strategy. Take along a gold bar and give it to your earlier self to keep until he or she embarks on the time travel; then there will be two gold bars. You will have effortlessly doubled your investment. It's just as easy to duplicate gold bars as people this way.

Nothing prevents this process from being repeated again and again.

From the physicist's point of view, duplicating entities is very disturbing, for it violates all sorts of so-called conservation laws. Suppose the gold bar were replaced by an electrically charged particle? Then two electric charges would appear from one. This violates the law of conservation of electric charge.

Again, paradox is avoided by sticking to self-consistent loops. For example, a positively charged particle taken through a wormhole will leave its electric field threading the hole, giving the wormhole an effective positive charge at the entrance (in the future) and a negative charge at the exit (in the past). The negative charge precisely cancels the additional positive charge that the trip back in time has created.

HOW TO PLUCK KNOWLEDGE OUT OF THIN AIR

The most baffling of all the time travel paradoxes is illustrated by the following parable. A professor builds a time machine in the year 2005 and decides to go forward (no problem there) to 2010. When he arrives, he seeks out the university library and browses through the current journals. In the mathematics section he notices a splendid new theorem and jots down the details. Then he returns to 2005, summons a clever student, and outlines the theorem. The student goes away, tidies up the argument, writes a paper, and publishes it in a mathematics journal. It was, of course, in this very journal that the professor read the paper in 2010.

Once more, there is no contradiction: the story involves a self-consistent causal loop so, strictly speaking, it is not a paradox but simply a very weird state of affairs. Rather, the problem concerns the origin of information. Where, exactly, did the theorem come from? Not from the professor, for he merely read it in a journal. But not from the student either, since he copied it from the professor. It's as if the information about the theorem just came out of thin air.

This paradox has a familiar ring to it. Something-for-nothing scenarios have long been pursued by eccentric inventors in search of perpetual motion. All such machines fail for reasons related to the first and second laws of thermodynamics, which, roughly, state that you can never get out of a closed system more than you put in. Proposed perpetual motion machines always generate some waste heat through friction and other inefficiencies, and eventually grind to a halt. Entropy (waste heat) and information are closely related (technically, rising entropy is the same as falling information). So getting information free is, from the physicist's point of view, equivalent to heat flowing backward, from cold to hot, which we should surely regard as a miracle.

Time travel expert David Deutsch believes that information entering the universe from nowhere is tantamount to a miracle and therefore strikes at the very heart of the

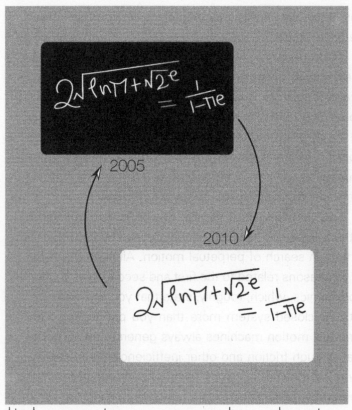

It is not a paradox but simply a very weird state of affairs.

orderly rationality of nature. For that reason he believes the third paradox is probably the most disturbing of the three. Perhaps we should place it alongside perpetual motion and cosmic censorship on our list of paradoxes, since all involve uncaused information entering the universe "from nowhere."

$\mathcal{O}$ How to Make Another Universe

At the heart of time travel paradoxes is the problem of causality: what happened yesterday affects what happens today. Go back and try to change yesterday and you threaten to change today too, making causal loops inherently problematic. But maybe there's a more comprehensive escape clause than restricting all time loops to detailed self-consistency.

Causality is not quite the rigid linkage that most people suppose. It's true that in daily life the connection between cause and effect is inescapable. However, the familiar world of tables and chairs and human beings conceals the shadowy microrealm of quantum mechanics in which causation is somewhat fuzzy (see p. 73).

The game of billiards provides a good example of commonsense causation at work. Hit the cue ball at a certain velocity so it collides with another ball. In the absence of causal loops, the motion of the two balls after the collision is completely determined by the initial speed and direction of the cue ball. Using Newton's laws of motion, you could work out in advance what will happen after the collision, because those laws are strictly *deterministic*—the initial state suffices to determine completely the final state. That is, if the experiment is repeated under identical conditions, the outcome should be exactly the same. If the struck ball drops into a particular pocket today, it will do so tomorrow, all else being equal. Thus is the orderly operation of the macrocosmos ensured.

Things are very different, however, if you try playing billiards with atoms, or particles like electrons and protons. Today, an electron may collide with a proton and bounce to the left. Tomorrow, *under identical conditions*, it may bounce to the right. Newton's laws of motion don't apply here and must be replaced by the rules of quantum mechanics, which are *indeterministic*. That is to say, the state of a physical system at one moment will not usually suffice to determine what will happen at the next moment. The uncertainty of the microrealm is encapsulated in Heisenberg's uncertainty principle (see p. 73). So prediction is a hazardous business in atomic theory. Generally the best that can be done is to give the betting odds for this or that outcome. An electron colliding with a proton might bounce off at one among a whole range of angles, some more likely than others. Quantum mechanics gives an accurate account of the probabilities, but it usually won't tell you what will happen in any given case.

Physicists are convinced that quantum uncertainty is

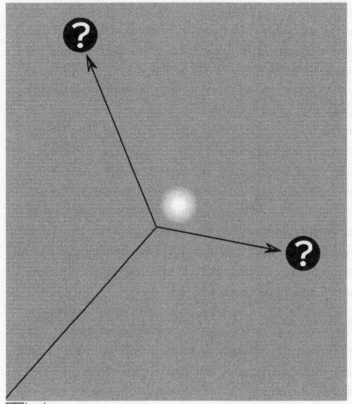

Things are very
different, however, if
you try playing billiards
with atoms.

intrinsic to nature, and not just the result of human igno-rance of the processes involved. In other words, even the electron doesn't know which way it will bounce until the collision actually happens. So although it remains true to say in a general sense that a collision with a proton causes the electron to be deflected from its path, the causal link is rather nebulous because the actual final path of the electron is undetermined.

It is not only simple collisions that are uncertain in atomic physics, but all processes. For example, a nucleus of the radioactive element uranium may or may not decay next year. An atom hitting a barrier may bounce back, or it may appear on the far side, having mysteriously tunneled through the barrier, because it is uncertain about precisely where it should be at any moment.

Among atoms and subatomic particles, quantum un-certainty is very conspicuous. However, for larger systems the fuzziness is less severe. When it comes to big mole-cules, quantum effects are rarely very important. But quantum uncertainty never completely disappears; in prin-ciple, it applies even to billiard balls.

If events in the microworld aren't completely nailed down by cause and effect, the whole complexion of the causal loop paradoxes associated with time travel changes. One way to think about quantum uncertainty is in terms of possible worlds. An electron hits a proton and may bounce either to the left or to the right. There are then two possible worlds: one with a left-moving electron, the other with a right-moving electron. More generally, an atomic or subatomic process will have many possible outcomes—perhaps even an infinite number of them—so there will be many alternative universes on offer almost every time something happens to a subatomic particle.

The issue of quantum uncertainty then forces itself upon us if we want to ask in any particular case: Which of the many possible universes will correspond to the *actual* universe? Of course, we cannot know in advance—that being the nature of quantum uncertainty—but most people suppose that there can be only one *real* world, all the others representing failed potential worlds. If that is so, there is then a deep problem about connecting smoothly between the quantum realm of multiple potential realities and the so-called classical (or everyday) realm of just a single reality.

In fact, there is no agreement about how to make this connection, but a growing number of physicists believe the best way to approach the problem is to suppose that each of those alternative universes is every bit as real as the others. In other words, there is no need to make a transition from many possible worlds to one actual world, because all the possible quantum worlds really exist. In this "many-universes" interpretation of quantum mechanics, there is an infinity of parallel universes, with each possible quantum alternative represented in a universe somewhere. There will be universes in which some atoms of your body are located in slightly different places, a universe in which President Kennedy wasn't assassinated, others that have no planet Earth, and so on. Every possible universe will be there somewhere, except that "there" means not out in space, but in some sense "alongside" our space and time (hence "parallel" universes). Zillions of these other universes will have copies of you, each of whom feels unique and assumes that he or she inhabits the one true reality.

Solving the time travel paradoxes by invoking parallel realities has long been a device used by writers of science fiction. The basic idea is that when the time traveler interferes with history, the universe forks into two or more branches. Among scientists who propose this escape route is David Deutsch, who points out that the many-universes interpretation of quantum mechanics naturally resolves the time travel paradoxes. Take the matricide paradox. Suppose the time traveler goes back and carries out the murder. This time there is no mistake; mother is dead. But which mother? Remember, there is a vast collection of mothers amid the stupendous number of parallel realities. In the multiverse of parallel quantum worlds, you could change the past of a parallel world, while leaving your own world untouched. In effect, the act of murder divides reality into two sets, one with a dead mother, another with a live mother. Both possibilities coexist side by side in the vastness of the quantum multiverse. Any given "branch" of the multiverse (that is, any specific observed reality) is thoroughly self-consistent, but causal interac-

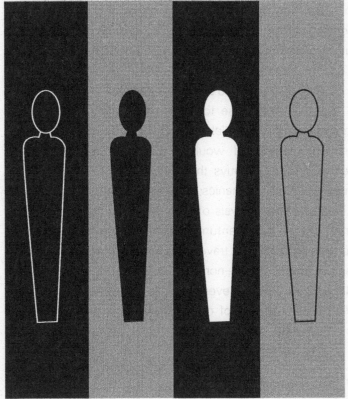

Copies of you, each of whom feels unique and assumes that he or she inhabits the one true reality.

tions between branches need not respect a definite chronological order. In Deutsch's scheme you can have your cake and eat it: time travel *and* unfettered free will are apparently both allowed.

Scientists are divided about the desirability of invoking the quantum multiverse to solve time travel paradoxes. Some believe that parallel realities are even more absurd than time loops and would rather have neither. But whether or not one buys the many-universes interpretation of quantum mechanics, nature *is* quantum mechanical, and any final analysis of a physical situation must be carried out at the quantum level. It seems that causal loops arising from time travel have the effect of amplifying quantum phenomena—normally confined to the atomic realm—to the level of everyday life. So we cannot avoid adding the weirdness of quantum reality to the strangeness of time travel.

Chronology Protection

Time travel may seem like fun to fans of science fiction, but the idea is positively scary for many physicists. The problem lies partly with the plethora of paradoxes that travel into the past would unleash. In addition, the possibility that causal loops may be imminent seems to be so physically pathological that it would produce profound physical effects; so profound, in fact, that they may stymie any attempt to actually create a time machine. Among theorists who have expressed serious doubts that wormholes or other time machines would work as advertised is Stephen Hawking. He has proposed a "chronology protection conjecture," which, in simplified terms, says that nature always comes up with an obstacle to prevent travel backward in time; "making the universe safe for historians" was the way he expressed it.

What might go wrong, then, if a supercivilization attempted to build a wormhole time machine? One possibility is that antigravity is too fickle a phenomenon to harness in realistic wormhole scenarios. It is one thing to demonstrate that negative energy is physically possible under some unusual circumstances, quite another to expect it to arise inside a wormhole or some other time machine setup, with the strength necessary to achieve time travel. The jury remains out on this one. Mathematical studies suggest that antigravity states in quantum fields occur under a fairly wide range of circumstances, but at present there is no general theorem to indicate precisely what the limits are.

Even granted that antigravity could be deployed in some suitable manner (or the necessary exotic matter obligingly put there by nature), other problems loom. The exotic matter pervading the wormhole's throat might interact with any normal matter attempting to traverse the wormhole and impede or destroy it.

Another difficulty concerns the behavior of the quantum vacuum in the vicinity of a wormhole, or any other sort of time machine. The problem centers on what happens at the join between the region of spacetime that permits time loops and "normal" spacetime where past and future don't get entangled. The interface between the two regions is called the *chronology horizon*. Crossing the chronology horizon entails entering a region of spacetime where particles can go round and round in endless causal loops. This includes the virtual photons of the quantum vacuum state. Crudely speaking, each time a virtual photon does a circuit in time, it doubles the (borrowed) energy. Calculations show that as the horizon is approached, virtual photons circulate around almost closed causal loops, and the nearer to the horizon they get, the closer to closure the loops risk becoming. Given the inherent uncertainties in the behavior of quantum particles like photons, the horizon does not act as a sharp boundary. The mere threat of impending causal closure is enough to boost the virtual photons without limit, piling up more and more energy as the horizon is approached. This runaway energy escalation would probably generate a huge gravitational field that would warp the spacetime and wreck the time machine. I say probably, because we do not yet have a good enough theory of quantum gravity to check what would actually happen under these extreme circumstances. So the quantum-vacuum-catastrophe argument is suggestive, but so far not fatal. At the time of writing, the chronology protection conjecture remains in limbo, a hope for some, a party pooper for others.

◯ Alternative Models of Time Machines

The wormhole remains the favored time machine design, but it is by no means the only one. I have already mentioned the early work of van Stockum and Gödel on rotating matter (see p. 33). A quite different proposal for a time machine has been made by J. Richard Gott III, based on the use of hypothetical entities called *cosmic strings*. A cosmic string is an astronomically long thread that contains a vast amount of mass; each kilometer of cosmic string would weigh about the same as the Earth. Some cosmologists believe cosmic strings may have formed in the hot big bang, when the intense primordial energy pervading space became trapped inside thin tubes and was preserved for posterity.

Cosmic strings would be made of exotic matter, but in this case what makes string matter exotic concerns not energy but *pressure*. Normally we do not notice that pressure is a source of gravity, but according to Einstein's general theory of relativity, pressure creates a gravitational field too. If it is truly enormous, pressure can rival energy in its gravitational power. It turns out that the pressure inside a cosmic string is enormous, and *negative*, which is to say that the string is in tension. Because pressure gravitates, tension (negative pressure) antigravitates. In the case of a straight segment of string, the antigravity of the tension exactly cancels the gravity of the mass-energy, with the result that the string would exert no gravitational force on a nearby body, in spite of its colossal mass.

Nevertheless, the string still alters the geometry of space in its vicinity, in a rather distinctive manner, best illustrated by analogy with a maypole. When a May Day dancer cavorts once around a maypole, he will turn through exactly 360 degrees. If the maypole were a cosmic string instead, the dancer would find he got back to his starting point after turning through less than 360 degrees. A circle drawn around a cosmic string does not contain four right angles as does a circle drawn on a blackboard.

The angular deficit caused by a cosmic string is predicted to be only a few seconds of arc, but, nevertheless, it leads to some distinctive effects. For example, a pair of parallel straight lines that straddle the string will end up converging. If the lines represent light rays from, for example, a quasar or distant galaxy, an observer will see two copies of either if the string interposes itself between it and the observer. Double images of this sort are known to astronomers, but they can be produced in other ways too, and there is no hard evidence that cosmic strings actually exist.

In spite of this, they are much studied. Gott has pointed out that photons from a distant source that strad-

Double quasar images
of this sort are known.

dle the string and converge need not arrive at the crossover point at the same moment if the string, source, and observer are not precisely aligned or are in relative motion. As a result, it would be possible for an astronaut traveling very close to the speed of light around one side of the string to reach the convergence point ahead of the photon coming around the other way. In effect, the astronaut will have outpaced the slower light pulse by taking an alternative route through space, just as in the case of the wormhole. This physical argument suggests that time travel may be possible using cosmic strings. Gott proved mathematically that if a pair of parallel cosmic strings are moving apart at very close to the speed of light, there will exist a region in which an astronaut could travel back in time by executing a loop around the strings.

Gott's proposal is hardly a practical one and encounters a number of objections on physical grounds; for example, infinitely long straight strings do not exist, while finite string loops threaten to collapse into black holes before they can be turned into time machines. But it establishes that time travel is a generic feature of Einstein's theory of relativity and not just a quirk of one scenario.

Gott has become so enthusiastic about time machines that he even goes so far as to suggest the entire cosmos may be one, pointing out that the universe would then be able to create itself. Just as a time traveler could, in principle, go back and become his own father (or her own mother), so the universe could loop back in time and bring itself into existence in a big bang without the need for a mysterious origin from nothing. In that way the universe will in some sense always have existed, even though time itself remains finite in the past.

This book by no means describes all possibilities for the design of time machines. Most proposals involve some sort of "cutting-and-pasting job" on spacetime, as if a superbeing wielding vast scissors hacks holes out of space, yanks and twists the exposed edges around, and then glues them together again in a different pattern. Although these schemes are very artificial, they all describe possible spacetimes, and they serve as test beds for exploring the amazing physical consequences of time travel.

⟂ Reversing Time

Time travel must not be confused with the equally fascinating (and equally speculative) topic of time reversal. Since at least the time of Plato, philosophers and scientists have mused about the idea of time "running backward." Actually this is a misnomer, since time itself doesn't run anywhere. It is more accurate to talk of physical systems running backward in time, like a movie played in reverse. Could this happen? Could water flow uphill or broken eggs reassemble themselves?

To get some idea of what is invloved, imagine a rigid box confining a dozen molecules of gas rushing about chaotically, colliding with each other and the walls of the box. Suppose at a certain moment all the molecules were crowded together in one corner. This arrangement wouldn't last long, as the speeding molecules bounced and scattered each other across the available space. The transition from "crowded" to "distributed" provides an "arrow of time" that serves to distinguish past from future. The existence of many such transitions in the world about us gives the impression that time has a definite direction associated with it. Time reversal would then entail such events as the rushing together of molecules of gas into one corner of a box. Is such a thing credible? It certainly is. One would expect that after a long enough duration, a dozen randomly moving molecules might find themselves revisiting one corner of the box together, purely by chance. In fact, it can be proved mathematically that such recurrences *must* happen repeatedly.

Of cource it's one thing for a handful of molecules to "go backward," quite another for the entire universe to reverse its normal behavior. The wait needed for things to get back to their starting arrangment grows rapidly with each extra particle involved. A typical room contains more than a trillion trillion molecules of air, which would take vastly longer than the age of the universe to congregate spontaneously in one corner, so there is no need to worry about suddenly being left gasping for breath. What this means is that while in principle the world could return to the state it had in, say, 1900, it is *exceedingly* unlikely to do so in our lifetimes, unless there's some built-in cosmic conspiracy among the countless particles.

Some physicists have conjectured that there may be just such a conspiracy programmed into the initial conditions of the universe, which would compel the entire cosmos to eventually revisit its starting condition in the big bang. We probably wouldn't know it if all the particles in the universe had been cleverly programmed to find their way back one day, thereby re-creating a past state. If this bizarre reversal were to happen, it would differ from time travel of the sort I have been discussing in this book in a fundamental way. Time reversal means *re-creating* the past, not revisiting it. If the universe did run backward, so too would human brain activity. We wouldn't see the great cosmic movie going in reverse, with stars sucking in light and black holes spewing out gas, because our minds and senses would be in reverse gear too. In short, living in a universe in which "time runs backward" would be no different from living in the one we now observe.

Why Study Time Travel?

The subject has provided fertile soil for novelists over the last century, cropping up repeatedly in both mainstream and science fiction. It has also provoked an extended (and rather confused) debate among philosophers about the nature of time and the logical contradictions that seem to occur when travel into the past is considered. Mostly, however, professional scientists have given the subject a wide berth—until recently. Now, research into time travel has become something of a cottage industry in the theoretical physics community. Some people find this surprising. We have seen how it still seems rather fanciful, drawing upon extremely speculative ideas of wormholes, cosmic engineering, and exotic forms of matter. How can professional scientists justify spending valuable time and research funds on such a frivolous topic?

Of course, there is no denying that it is fun, and that some scientists treat the subject as an intellectual game. But there is a serious side to it too. The "thought experiment" is a time-honored part of the scientific process. It works by the scientist's dreaming up a scenario, which may appear at the time to be fantastical, in order to push current theories to their outer limits. The purpose in so doing is to expose any logical flaws or inconsistencies in the theory. Thought experiments enabled Galileo to deduce the law of falling bodies by pure reasoning alone. They also led Einstein to correctly predict the time dilation effect. In the 1930s, thought experiments such as the one associated with the famous Schrödinger's cat paradox played an important role in clarifying the meaning of quantum mechanics. Significantly, advances in technology have now enabled many thought experiments to be performed as real experiments.

Just the fact that time travel seems doubtful, or even impossible to us today, doesn't mean that we can ignore its implications. It may be that easier ways to build a time machine will be discovered, ways that would not require the resources of a supercivilization. But the very possibility of visiting or signaling the past presents a serious challenge to our understanding of physics, regardless of whether or not time travel ever becomes a practical proposition. Researchers agree that any attempt to make a time machine would almost certainly generate dramatic quantum vacuum effects, the consequences of which cannot be fully explored without a tractable and reliable theory of quantum gravity. Since achieving such a theory is currently a priority among theoretical physicists, the study of time loops and the resulting illumination of the deep causal structure of the universe is very timely, so to speak.

Part of the fascination of time travel concerns the stark paradoxes that threaten as soon as travel into the past is considered. The purpose of science is to provide a consistent picture of reality, so if a scientific theory produces genuinely paradoxical (rather than merely weird or counterintuitive) predictions, that is a very good reason for rejecting the theory. As we have seen, time travel is replete with paradoxes. At the moment, opinions differ markedly on how to deal with them. Perhaps causal loops can be made self-consistent. Perhaps reality consists of multiple universes. Or maybe our description of nature must be radically revised. On the other hand, there may be no way to evade the paradoxical nature of time travel, and we shall be obliged to invoke Hawking's chronology protection conjecture (see p. 113) and discard all theories that permit travel into the past.

Most recent attempts to provide a quantum description of gravity are formulated within the broader context of a completely unified theory of physics, in which all the particles and forces of nature, along with space and time, are amalgamated in a common mathematical scheme. Fashionable among these "theories of everything" are superstrings, and the more comprehensive scheme known cryptically as M-theory.

It is fascinating to speculate that chronology protection could be a global principle of nature on a par with, say, the second law of thermodynamics. We might even compile a list of cosmic taboos—

No time machines!
No perpetual motion machines!
No naked singularities!
Etc.

—and use this list as a filter for physical theories. Any theory that did not respect all the taboos on the list should be rejected. That would be an excellent way of culling contender theories. If the list is long enough, it may happen that only one "theory of everything" would pass through the filter. We would then know the answer to the ultimate scientific question: Why *this* universe rather than some other?

Bibliography

NONFICTION

Al-Khalili, Jim. *Black Holes, Wormholes & Time Machines*. Bristol, U.K.: Institute of Physics Publishing, 1999. A good, clear introduction to relativity, cosmology, and gravitation, with a large section on time travel.

Berry, Adrian. *The Iron Sun*. London: Jonathan Cape, 1997. An early speculation about crossing the universe using a black hole/wormhole.

Davies, Paul. *About Time*. London: Penguin, 1995. An in-depth survey of the subject of time in its many aspects.

Deutsch, David. *The Fabric of Reality*. London: Penguin, 1997. An exposition of the many-universes interpretation of quantum mechanics, including its relevance for time travel.

Gott, J. Richard III. *Time Travel in Einstein's Universe*. Boston: Houghton Mifflin, 2001. A good technical summary of time travel, with special emphasis on the cosmic strings model.

Greene, Brian. *The Elegant Universe*. New York: Norton, 1999. A lucid account of recent attempts to unify the fundamental forces and particles of nature.

Nahin, Paul J. *Time Machines*. Woodbury, N.Y.: AIP Press, 1993. A fascinating survey of time travel in fiction and nonfiction. Many references.

Novikov, Igor D. *The River of Time*. Cambridge: Cambridge University Press, 1998. A very readable account of relativity. Includes a section on time travel.

Pickover, Clifford. *Time: A Traveller's Guide*. Oxford: Oxford University Press, 1999. A readable survey.

Thorne, Kip S. *Black Holes and Timewarps*. New York: Norton, 1994. An extensive, detailed account of the general theory of relativity, black holes, and wormholes by one of the key players. Many references to the original literature.

Wheeler, John A. *A Journey into Gravity and Spacetime*. New York:

Scientific American Library, 1990. From the man who coined the terms "black hole," "wormhole," "spacetime foam," and much else.

Will, Clifford. *Was Einstein Right?* New York: Basic Books, 1986. An excellent introduction to the theory of relativity and experimental tests thereof.

FICTION

Benford, Gregory. *Timescape.* New York: Simon & Schuster, 1980. Reissue, New York: Spectra, 1996. Written by a professional physicist, this award-winning science-fiction story includes this author as a character!

Benford, Gregory. *Cosm.* New York: Avon Eos, 1998. A hard sci-fi story about the creation of a baby universe in the laboratory, initiated by a heavy-ion collision at the Brookhaven National Laboratory.

Bradbury, Ray. "A Sound of Thunder." In *The Stories of Ray Bradbury.* New York: Knopf, 1980. Short story illustrating how the future depends delicately on small details of past states.

Crichton, Michael. *Timeline.* New York: Random House, 1999. Drawing upon ideas of quantum wormholes, Crichton weaves an action-packed time travel drama, with an attempt at a self-consistent history.

Sagan, Carl. *Contact.* New York: Simon & Schuster, 1985.

Wells, H. G. *The Time Machine and Other Stories.* London: Penguin, 1946. The classic, founding story by the master himself.

TECHNICAL

Gödel, K. "An Example of a New Type of Cosmological Solution of Einstein's Field Equations of Gravitation." *Reviews of Modern Physics* 21 (1949): 447–50.

Hawking, S. W. "The Chronology Protection Conjecture." *Physical Review D* 46 (1992): 603–11.

Morris, M. S., and Thorne, K. S. "Wormholes in Spacetime and Their Use for Interstellar Travel: A Tool for Teaching General Relativity." *American Journal of Physics* 56 (1988): 395–412.

Roman, T. A. "Inflating Lorentzian Wormholes." *Physical Review D* 47 (1993): 1370–81.

Tipler, F. J. "Rotating Cylinders and the Possibility of Global Causality Violation." *Physical Review D* 9 (1974): 2203–6.

Visser, Matt. *Lorentzian Wormholes: From Einstein to Hawking.* Woodbury, N.Y.: AIP Press, 1995.

Index

Eliot, T. S., 93
Euclidean geometry, 41

First Men in the Moon, The (Wells), 61
Flamm, Ludwig, xi, 45
Foster, Jodie, 58, 59, 60
free will, in time travel paradox, 96–97, 99
Fulling, Stephen, 82, 86
future, 2, 3
    backward time travel of tourists from, 93–94
    time travel into, 5–31

Galileo, 22, 51, 122
general theory of relativity, xi
    antigravity in, 62, 63
    backward time travel and, 32–34, 118
    gravitational force in, 14–22, 32–33, 38, 49–53, 60, 62, 63, 115
    spacetime warping in, 14–22, 34–35, 38
Gödel, Kurt, xi, 33–34, 115
Gott, J. Richard, III, xiii, 115–19
gravitational force:
    of accelerated particles, 23–25
    of black holes, 37–38, 43, 53–58, 61, 86–87
    centrifugal force vs., 53
    counteracting of, *see* antigravity
    of Earth, 15–17, 54, 85–86
    in general theory of relativity, 14–22, 32–33, 38, 49–53, 60, 62, 63, 115
    negative energy as by-product of, 85–86
    of neutron stars, 17–22, 36–38, 89–90
    of pressure, 115–16
    in spacetime, 49–53
    of Sun, 41–45
    of wormholes, 40–49, 53–58, 61

Hafele, Joe, 7
Haldane, J. B. S., 4
Hawking, Stephen, xii, 86–87, 113, 124
Higgs field, 79

imploder, quark-gluon plasma compressed by, 77–79

inflators, 79–88
    gravitational fields and, 85–88
    negative energy and, 80–85
Institute for Advanced Study, 33

Jupiter, 54

Keating, Richard, 7
Kruskal, Martin, 45

Large Electron Positron (LEP) collider, 12, 25
lasers, negative energy created by, 84–85, 87
Levine, Martin, 15
levitation, 61
light, speed of, 9–11, 23–25, 31
light barrier:
    accelerated particles and, 25, 98
    causality of time-ordered events and, 28–30, 34
    rotating time machines and, 34–35
    in special theory of relativity, 9–11, 23–25, 98

magnetic pinching, 77–78
Milky Way galaxy:
    black holes in, 36
    distances in, 11
M-theory, 124
multiverses, 108–12, 124
muons, 12

negative energy:
    antigravity and, 62–64
    of black holes, 86–87
    of Earth's gravitational field, 85–86
negative energy, creation of:
    Casimir effect in, 76, 80–82
    lasers used in, 84–85, 87
    "moving mirror" effect in, 80, 82–84
negative pressure, cosmic strings and, 115–16
neutron stars:
    black holes and, 36–38
    as differentiators, 89–90
    gravitational force of, 17–22, 37–38, 89–90
    spacetime warping of, 17–21, 36–38
    as time machines, 21, 30